Techniques in Microbiology
A Student Handbook

John M. Lammert
Gustavus Adolphus College

San Francisco Boston New York
Cape Town Hong Kong London Madrid Mexico City
Montreal Munich Paris Singapore Sydney Tokyo Toronto

Lammert, John
Techniques in microbiology : a student handbook / John M. Lammert.
p. ; cm.
Includes index.
ISBN 0-13-224011-4 (spiral bound : alk. paper)
1. Microbiology—Laboratory manuals. 2. Microbiology—Handbooks, manuals, etc. I. Title.
[DNLM: 1. Bacteriological Techniques—methods—Laboratory Manuals. QW 25.5.B2 L232 2007]
QR63.L355 2007
579.078—dc22
2006029624

Executive Editor: *Gary Carlson*
Project Manager: *Crissy Dudonis*
Editor-in-Chief, Science: *Dan Kaveney*
Production Editor: *Laserwords Private Limited*
Executive Managing Editor: *Kathleen Schiaparelli*
Assistant Managing Editor: *Beth Sweeten*
Manufacturing Manager: *Alexis Heydt-Long*
Manufacturing Buyer: *Alan Fischer*
Art Director: *Jonathan Boylan*
Interior Designer: *Koala Bear Design*
Cover Designer: *Jonathan Boylan*
Senior Managing Editor, Visual Asset Production and Management: *Patricia Burns*
Managing Editor, Art Management: *Abigail Bass*
Art Production Editor: *Thomas Benfatti*
Illustrations: *Laserwords Private Limited*
Cover Credit: Background image—*Red leather texture, copyright J. Helgason, Shutterstock, Inc.;* illustration—*Laserwords Private Limited*

Printed in the United States of America

36 2022

ISBN 0-13-224011-4

Pearson Education Ltd., *London*
Pearson Education Australia Pty. Limited, *Sydney*
Pearson Education Singapore, Pte. Ltd
Pearson Education North Asia Ltd., *Hong Kong*
Pearson Education Canada, Ltd., *Toronto*
Pearson Education de Mexico, S.A. de C.V.
Pearson Education—Japan, *Tokyo*
Pearson Education Malaysia, Pte. Ltd.

To Jennifer: wife, soul mate, and hand model extraordinaire
and to
Emily, Annika, and Hannah: daughters and the stars in my eyes

CONTENTS

Preface ix

About the Author xi

Unit I – Safety First! 1

Safety in the Microbiology Laboratory 1

Universal Human Blood and Body Fluid Precautions 5

Unit II – Culturing Bacteria and Aseptic Techniques 6

Preparing Culture Media 6

Sterilization 17

- The Autoclave 17
- Dry Heat 18
- Filtration 19

Adjusting the Gas Burner 23

Using an Electrical Incinerator (Bacti-Cinerator) 27

Aseptic Transfer of Bacteria 28

- Preparation of the Work Area 28
- Transferring from a Broth Culture into a Broth Tube 29
- Transferring from a Broth Culture onto an Agar Slant 32
- Transferring from an Agar Plate to an Agar Slant 35
- Transferring Bacteria while Holding Multiple Tubes 37
- Inoculating an Agar Deep Using the Stab Technique 40

Isolation of Bacteria 42

- Preparing a Streak Plate 42
- Preparing a Pour Plate 48
- Preparing a Spread Plate 56

Characteristic Features of Bacterial Growth in Culture 60

Maintenance and Storage of Stock Cultures 63

- Short-term Storage of Stock Cultures 63
- Short- and Long-term Storage 63
- Growing a Fresh Culture from Frozen Stock 65

Culturing Anaerobic Bacteria 70

Unit III – Visualizing Bacteria **74**
Effective Use and Responsible Care of the Light Microscope 74
Viewing Live Microorganisms 84
Preparing a Hanging-drop Slide 84
Staining Bacteria 87
Preparing a Bacterial Smear 87
Preparing a Simple Stain 92
Preparing a Gram Stain 94
Preparing an Acid-fast Stain 99
Preparing a Negative Stain 105
Preparing a Capsule Stain 109
Preparing an Endospore Stain 112
Measuring Microscopic Cells 117

Unit IV – Enzyme-based Tests for the Identification of Bacteria **121**
Hydrolytic (Digestive) Enzymes 121
Starch Hydrolysis Test 121
Casein Hydrolysis Test 123
Gelatin Hydrolysis Test 125
Fat (Triglyceride) Hydrolysis Test 127
DNA Hydrolysis Test 129
Utilization of Carbohydrates 131
Fermentation of Carbohydrates (Durham Tube) 131
Methyl Red Test (Mixed Fermentation) 133
Voges-Proskauer Test (Butanediol Fermentation) 135
Citrate Utilization Test 137
Oxidation–fermentation (OF) Glucose Test 139
Degradation of Amino Acids 142
Indole (Tryptophan Degradation) Test 142
Hydrogen Sulfide (H_2S) Production Test 144
Phenylalanine Deamination Test 146
Amino Acid Decarboxylase Test 148
Respiration Tests 150
Catalase Test 150
Oxidase Test 152
Nitrate Reduction Test 154

Miscellaneous Tests 157
Triple Sugar Iron (TSI) Agar Test 157
Urea Hydrolysis Test 159
Litmus Milk Reactions 161
Motility Testing 164
Coagulase (Tube) Test 166
Selective and/or Differential Media 168
Blood Agar Plate 168
Eosin Methylene Blue (EMB) Agar 170
Mannitol Salt Agar 172
MacConkey Agar 174
Phenylethyl Alcohol (PEA) Agar 176

Unit V – Counting Microbes 178

Direct Counting with the Petroff-Hausser Chamber 178
Preparing a Standard Plate Count of Viable Bacteria 183
Using Turbidimetry to Estimate Cell Density 191
Determining Bacteriophage Titer by a Plaque Assay 198

Unit VI – Measuring the Effectiveness of Antibacterial Chemicals 206

Testing Antibacterial Medicines: Kirby-Bauer Technique 206
Testing Antiseptics and Disinfectants: Disk Diffusion Technique 213

Appendix: Stains and Reagents 218

Index 223

PREFACE

This handbook has been prepared for you, a microbiology student, with concise instructions for carrying out the manipulations and procedures that are fundamental in the microbiology laboratory. These are the techniques that you will frequently and often repeatedly use as you study bacteria in the laboratory. To help you visualize the "how-to-do-it" aspect of these techniques, instructive graphics are included. Color artwork clearly illustrates each step in a procedure.

The laboratory techniques and procedures presented in this handbook have been compiled from recommendations made by microbiology instructors who teach at two-year, four-year, and research colleges and universities. And most importantly, these techniques include those skills considered essential for the introductory microbiology laboratory core curriculum, as collectively prepared by members of the education community in the American Society for Microbiology.

Even though most of the bacteria you will encounter in the instructional microbiology laboratory are not particularly good at infecting healthy people, some do have this potential. To protect you from unwanted intrusion of your body by these microscopic creatures, safety is emphasized in this handbook.

Where appropriate, safety issues that must be kept in mind when you carry out a particular technique are highlighted by the icon shown to the left.

To help you work through those inevitable glitches that happen the first time you perform an unfamiliar procedure, a Troubleshooting section is included for many procedures. You will find tips for preventing and overcoming these "bumps in the road."

You have probably noticed that this handbook is smaller than lab manuals that you have used in other courses. This convenient size is intentional. The handbook will take up less space on the lab bench top, reducing the chance that you might spill growth media with bacteria on its pages. It has been designed to minimize turning pages as you follow the concise text and graphics. And the compact size also means less weight in your backpack. If you happen to slip the handbook into your lab coat pocket, do not forget to take it with you when you have finished your work.

When you master the techniques presented in the six thematic units of this handbook, you will be competent in culturing and handling bacteria, visualizing these microscopic creatures, identifying bacteria on the basis of their staining characteristics and their complement of enzymes that allow them to utilize substrates in the environment, estimating their number (including bacterial viruses), and determining the effectiveness of antibacterial chemicals. You can then use these skills for more-in-depth studies of bacteria and bacteriophage.

This visual guide to microbiology laboratory techniques and procedures complements an in-house or commercial lab manual. It is also quite appropriate for a microbiology course with a lab component that focuses on discovery or is investigation oriented.

Acknowledgments

An author only provides the text and suggestions for artwork in a book. The real work for preparing a publication that is distinctly instructive and visually effective is accomplished by a competent publishing team. Many thanks to Gary Carlson, Executive Editor, the catalyst with the idea for the project and the bringing together of microbiology educators to focus on the idea; to Tom Benfatti, AV Editor, the translator from digital pictures to marvelous color artwork; to Jonathan Boylan, Art Director, the vision for the overall look of the handbook; to Erin Connaughton, Full Service Manager at Laserwords, the organizer for the final production. And an extra special thanks to Crissy Dudonis, Project Manager, patient guide, and hand-holder who worked diligently to keep me on task and to bring the handbook together.

I wish to extend thankful recognition to focus group members who met after an ASM Conference for Undergraduate Educators to critique the idea of this handbook. They include:

Mary Allen, *Hartwick College*
Gail Baker, *Okaloosa-Walton College*
Betsey Dyer, *Wheaton College*
Anne Hanson, *University of Maine*

I am also grateful to the comments and suggestions from reviewers:

Shivanthi Anandan, *Drexel University*
Elizabeth Ehrenfeld, *S. Maine Technical College*
Nitya Jacob, *Emory University*
James Martiney, *Pace University*
Kristine Snow, *Fox Valley Technical College*
Neil Baker, *Ohio State University*
Michelle Furlong, *Clayton State University*
Denise Poteat, *Ohlone College*
Jackie Reynolds, *Richland College*
Luanne Wolfgram, *Johnson County Community College*

If you find some step in a technique is incorrectly stated, if you think of a more effective presentation of a procedure, or if you wish an additional technique had been included, please contact me. I continue to enjoy conversations with other microbiology educators and to delight in learning new ways to facilitate learning by my students.

John M. Lammert
Gustavus Adolphus College
jlammert@gustavus.edu

About the Author

John M. Lammert, Associate Professor in the Department of Biology at Gustavus Adolphus College, St. Peter, MN, teaches "Microbes and Human Health," "Microbiology," "Immunology," "Principles of Biology," and "First-term Seminar: Viruses." He has also been on the faculty at Valparaiso University (St. Louis division) and Miami University. Dr. Lammert received a B.A. (mathematics) and an M.A. (biology) from Valparaiso University, and a Ph.D. (immunology) from the University of Illinois-Medical Center, Chicago. He has served the membership of the American Society for Microbiology as the first chair of Division W (Microbiology Education), as a member of the Council as Counciler-at-large, as a member of the Undergraduate Education Committee, and as a member of several planning committees for ASM Conferences for Undergraduate Educators. Dr. Lammert has received the Edgar M. Carlson Award for Distinguished Teaching at Gustavus. He has also written three books on science fair projects (microbes, plants, and the human body).

At the annual Nobel Conference held at Gustavus, Dr. Lammert has been host for some of his "scientific heroes," including Nobel laureates Christian Anfinsen, Salvador Luria, Baruj Benacerraf, and Günter Bloebel, as well as Robert Gallo and Leonard Hayflick. Dr. Gallo was also co-instructor with Dr. Lammert in a Spring 2006 course "Biology of Viruses."

SAFETY IN THE MICROBIOLOGY LABORATORY

Follow these safety rules and practices to:

- reduce significantly or eliminate the risk of accidental infection in the microbiology laboratory.
- prevent potentially harmful microbes from leaving the laboratory.

Protect yourself and others

- Wash your hands with antibacterial soap and water in the lab before you begin and after you complete your work, as well as before you leave the lab for any reason. Liquid or powdered soap from a dispenser is preferred: bar soap is likely to become contaminated.

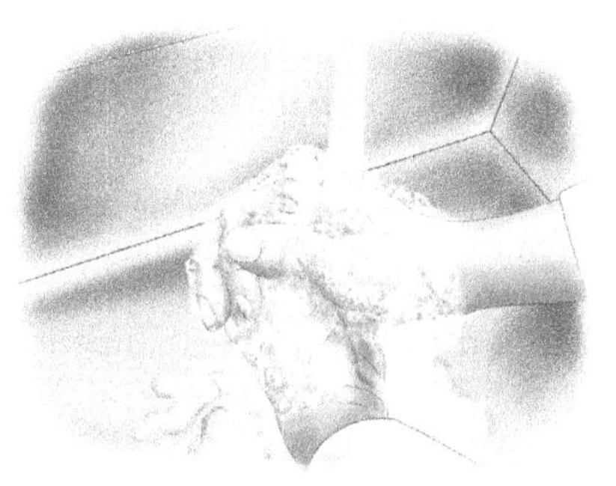

- Decontaminate the work surface with a disinfectant *before* you begin and *after* you complete lab work.

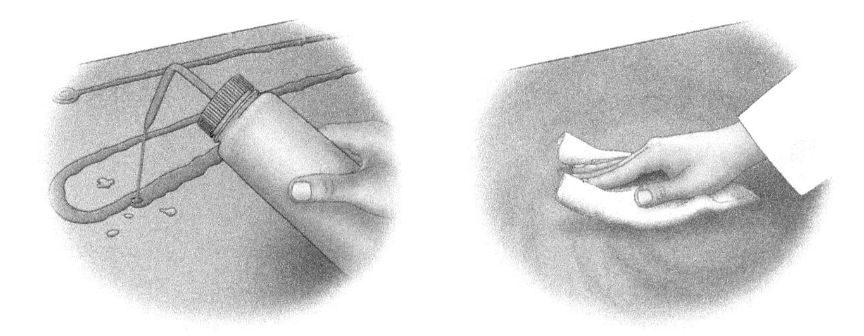

- Leave your coat, hat, and backpacks outside the lab or place them out of the way in a cupboard. Keep only your lab manual and/or notebook on the lab bench top.
- Wear only closed-toed shoes in the laboratory. Do not wear flip-flops, sandals, or any open-toed shoes.
- Wear protective clothing, preferably a lab coat. Your lab coat should never leave the lab. If it is made of cloth, wash it periodically in hot water with bleach and detergent.
- Wear safety glasses except when you are viewing slides under the microscope.
- Wear gloves if you have open lesions or a rash on one or both hands, or if your immune system is compromised. If you or anyone in the lab is allergic to latex, use gloves made of nitrile. Wash your hands with soap and water after you remove the gloves.
- When you are using a gas burner (or other open flame), position it so that you will not burn yourself. Tie long hair back to prevent it catching on fire. Turn off the gas when the flame is not needed.
- Keep your hands away from your mouth and eyes. This means no handling contact lenses, no applying cosmetics (including lip balm), no nibbling your fingernails or a pencil, and no rubbing your eyes.
- Do not eat or drink while you are in the lab. Food and beverages must remain outside the room.
- Avoid the use of electronic devices, including cell phones, laptops, MP3 players, and PDAs. If your device were to become contaminated, disinfecting it could cause damage.
- Never use your mouth to draw liquids into a pipette or tube: use a mechanical pipetting device.

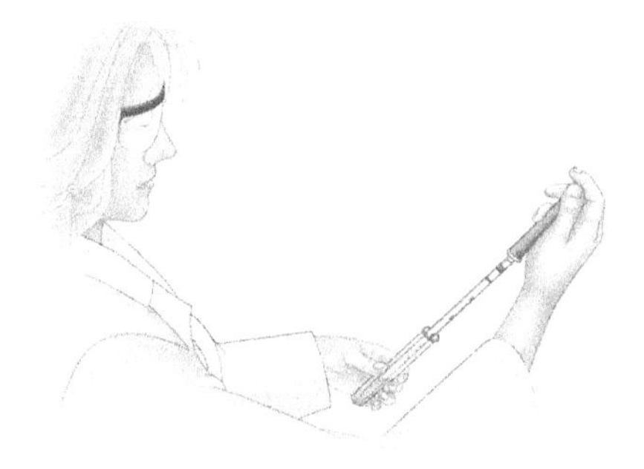

- Keep cultures in a test tube rack or other holder whether you are working at your bench or walking about in the lab.

- Place all contaminated tubes, plates, and waste materials in appropriate receptacles for subsequent sterilization.

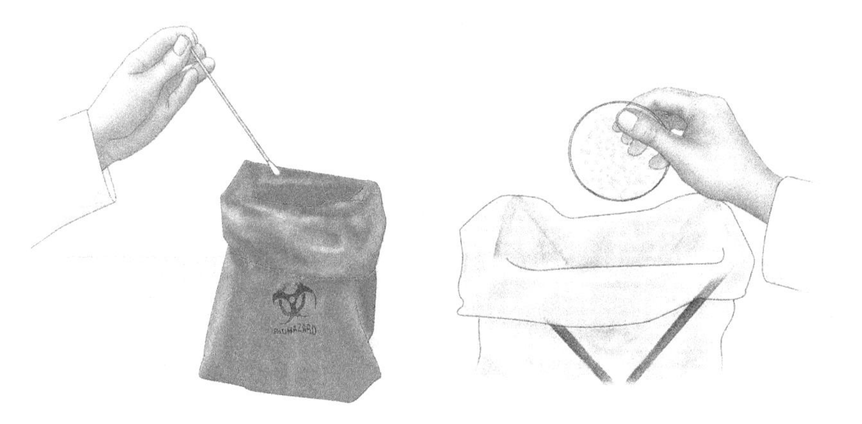

- If you spill a chemical on your skin, immediately wash the affected area. If a chemical or a culture gets into your eye, immediately wash your eyes for 5 minutes.
- If anything containing viable bacteria spills, cover the contaminated culture media and/or broken glass with a paper towel and a generous amount of disinfectant. Allow the disinfectant to remain on the spill for at least 20 minutes before cleaning it up.
- Sweep up pieces of broken glass (after disinfection if necessary) and place them in the specified container. Never use your hands to pick up the pieces.

- Place sharps, such as scalpel blades or syringes with needles, in a designated sharps container.

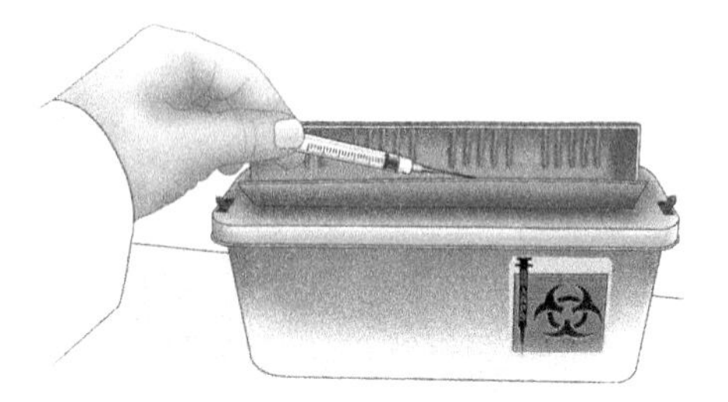

- Do not keep cultures in your lab drawer or locker when you have finished with them.
- Do not remove cultures, reagents, or other materials from the lab unless you have permission to do so.
- Post a biohazard sign on the door to the lab to alert visitors to possible risks.
- Know the locations of the first aid kit, the eye wash station, and the fire extinguisher. Know the fire emergency route out of the building.

Universal Human Blood and Body Fluid Precautions

If you handle human blood or any other body fluid, you must handle these fluids as though they contain blood-borne pathogenic microbes: for example, human immunodeficiency virus (HIV), hepatitis B virus (HBV), and hepatitis C virus (HCV). These precautions must be taken in addition to those biosafety criteria listed in "Safety in the Microbiology Laboratory."

- Always wear gloves. If you or any one in the lab is allergic to latex (powder from latex gloves can elicit a severe allergic reaction in susceptible people), wear gloves made of nitrile.
- Change gloves immediately if they become torn.
- Remove gloves before you touch any lab equipment.
- Never touch your face.
- Wear a lab coat. If it becomes soiled or spattered with body fluid, it must be washed in hot water, detergent, and bleach. Once used, the lab coat is never worn outside the lab until it is washed.
- Wear safety goggles or a face mask for protection against possible spattering.
- Dispose of gloves or any other material that has come into contact with blood or other body fluids in a biohazard container for sterilization.
- After you take off your gloves, wash your hands immediately for at least 15 seconds with antibacterial soap and water.
- To disinfect spills of blood or other body fluids, use a germicide approved by the EPA or diluted household bleach (sodium hypochlorite). Prepare fresh dilutions of bleach for each spill: 1:100 for smooth surfaces or 1:10 for dirty or porous surfaces. Allow the disinfectant to remain on the spill for at least 20 minutes, then wipe up with disposable towels. Discard the towels in a biohazard container for sterilization.

PREPARING CULTURE MEDIA

Purpose:

- To prepare sterile media that provide nutritional requirements for cultivation of bacteria.

Materials:

- dehydrated medium in bottle
- balance or scale
- weighing boats or weighing paper
- spatula
- stir plate with heater
- magnetic stir bar
- pH meter or pH paper
- 1 N HCl and 1 N NaOH
- deionized or distilled water
- large glass flask or beaker
- glass test tubes (16 × 150 mm)
- enclosures for test tube mouths (e.g., slip-on caps, screw caps, or cotton plugs)
- test tube rack or basket
- Petri plates (plastic or glass)

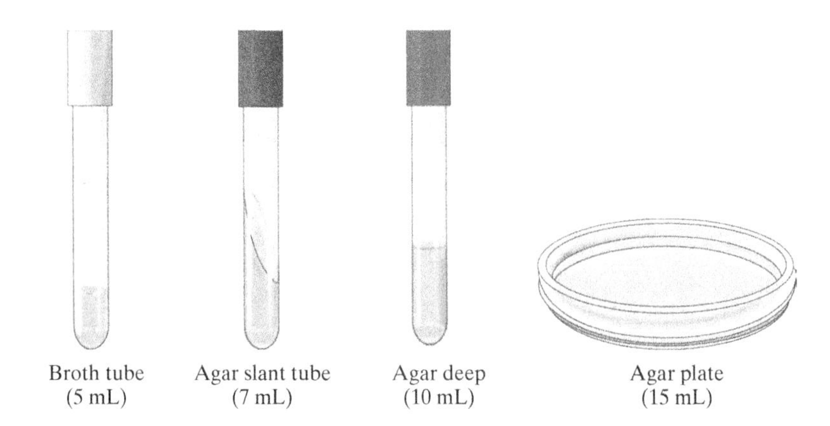

Types of containers with culture media

Preparing a batch of medium

1. Weigh out the correct amount of dehydrated medium.

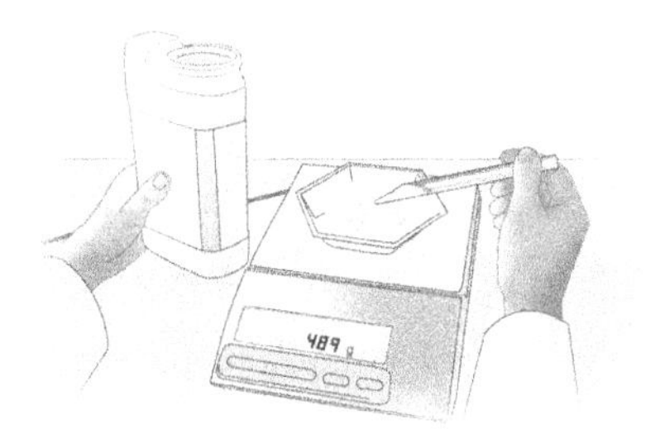

2. Add the weighed dehydrated medium to a measured volume of deionized or distilled water. The volume of the beaker or flask should be at least twice that of the volume of water used.

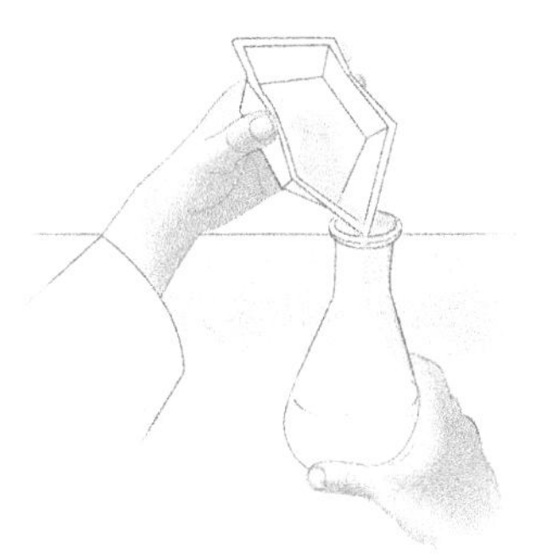

3. Adjust the pH by adding drops of 1 N HCl or 1 N NaOH, as stated on the medium bottle. Use pH paper or a pH meter.

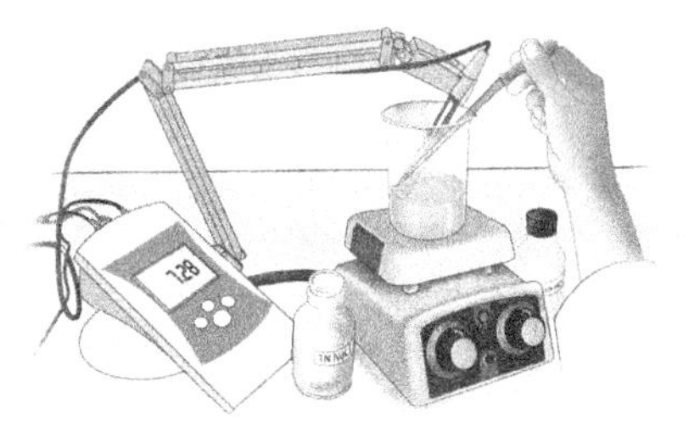

Preparing broth tubes

1. After the dehydrated medium has dissolved, dispense the appropriate volume into the test tubes. (The method for a large number of tubes is shown.)

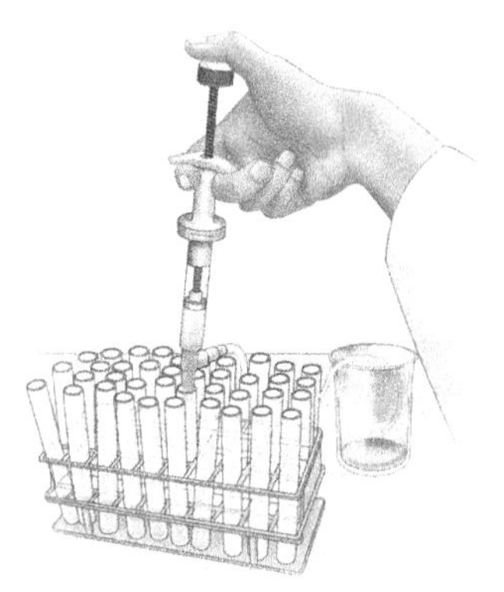

1a. Use a pipettor and a sterile pipette to dispense medium into a small number of test tubes.

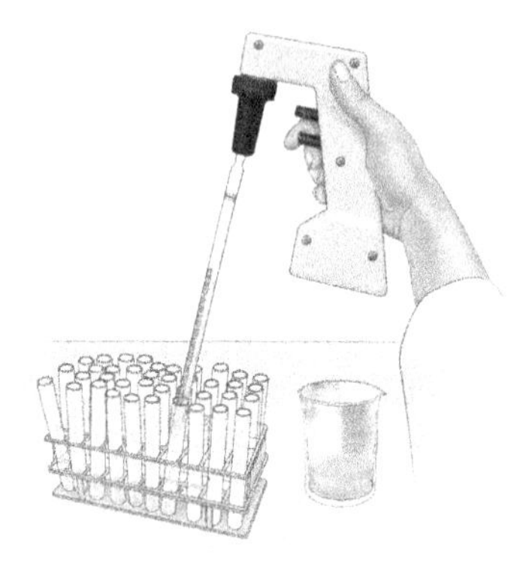

2. Place caps on the filled test tubes.

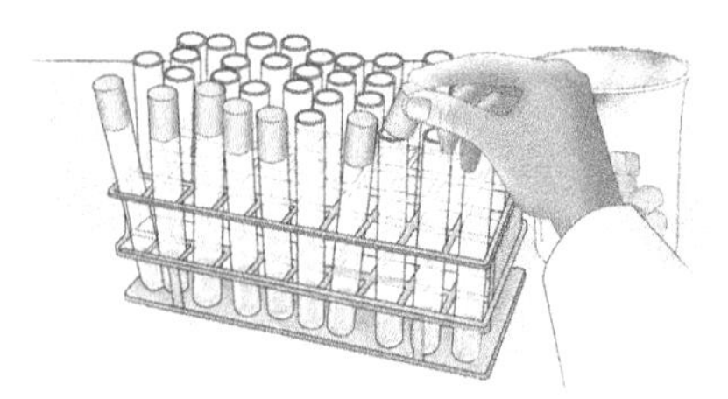

2a. For screw caps, leave the caps partially unscrewed to allow steam to enter and escape.

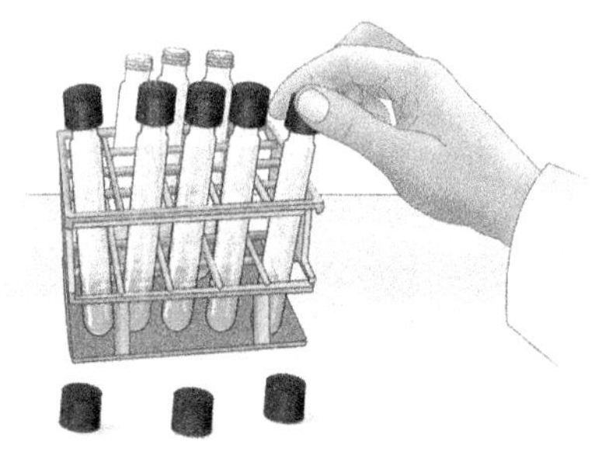

3. Label the test tube rack or basket with the name of the medium and the date.

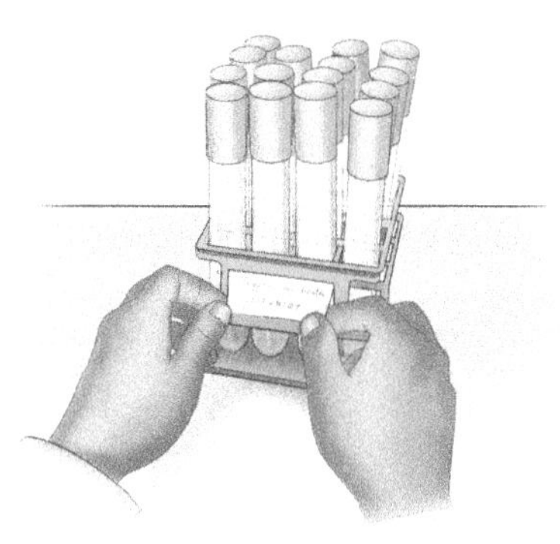

4. Sterilize the test tubes in the autoclave.

5. After sterilization is complete, remove the rack or basket. You must wear a protective glove when you handle a rack or basket with test tubes after autoclaving: they are very hot.

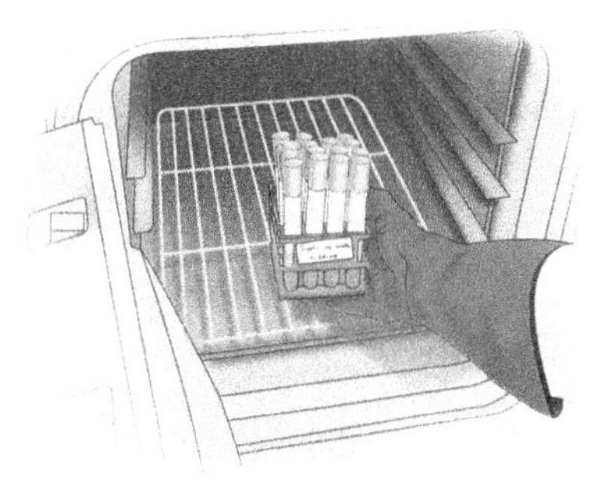

6. Cool the test tubes at room temperature. Store them at 4° C.

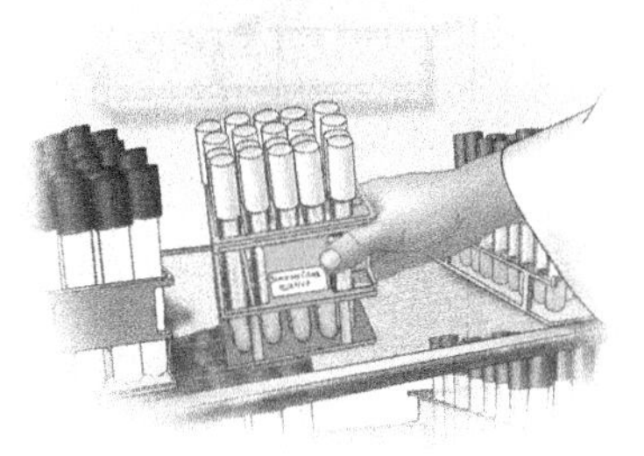

Preparing agar deep tubes and agar slant tubes

Procedure:

1. Melt the agar medium by boiling it in a flask that holds at least twice the volume of the medium. Use a magnetic stirrer and stir bar to mix it.

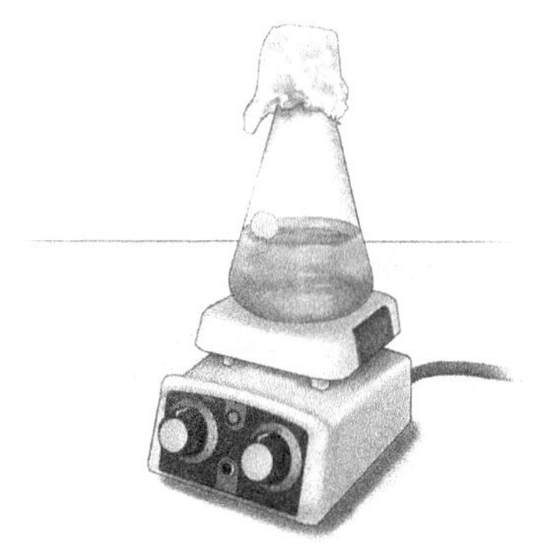

2. After about 15 minutes of cooling at room temperature, dispense the appropriate volume of melted agar into each test tube. Cap the tubes.

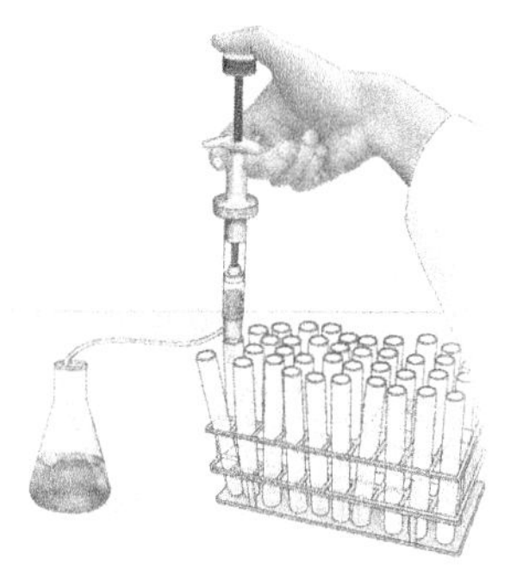

3. Label the test tube rack or basket with the name of the medium and the date. Sterilize the tubes in the autoclave.

4. To prepare agar slant tubes, on the lab bench, tilt the tubes of liquid agar (cap end up) on a board that is about $^1/_2$ inches thick.

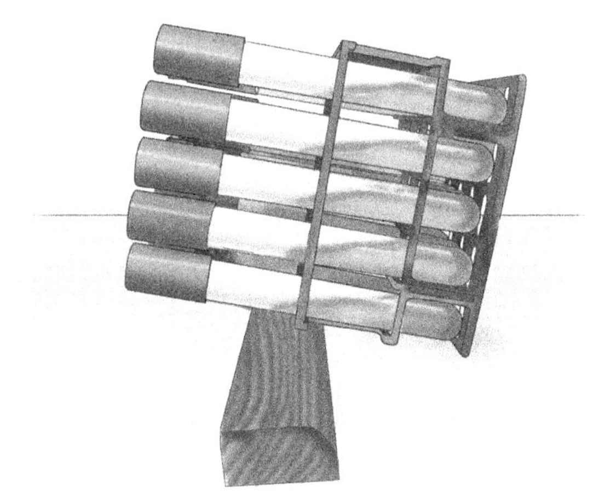

5. After the agar has solidified (about 25 minutes), store the tubes at 4° C.

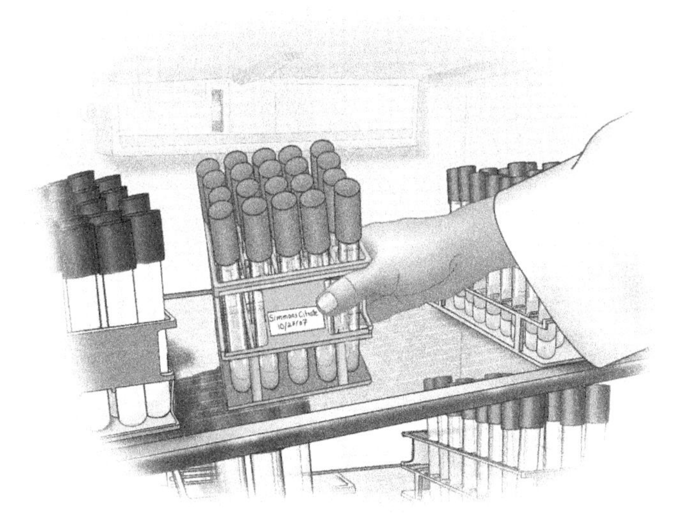

Preparing agar plates

Method I: Pouring sterile melted agar from tubes into plates

Procedure:

1. Prepare melted agar medium as described above. Dispense 15 mL of melted agar into each tube. Set the tubes in a test tube rack and sterilize them in the autoclave.

2. After autoclaving, place the rack and tubes in a 48° C water bath. You must wear a protective glove when you remove the tubes from the autoclave.

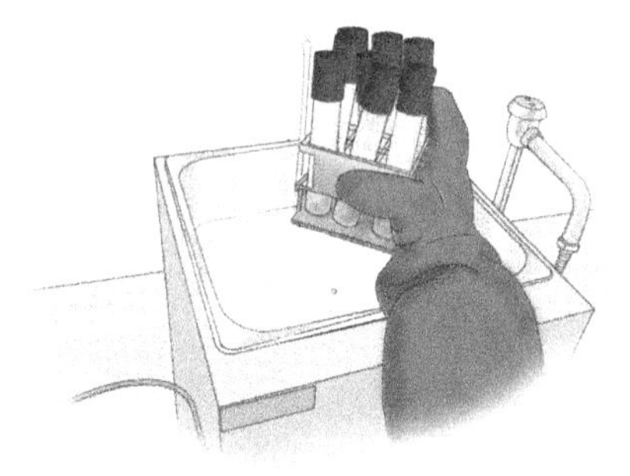

3. Place sterile Petri plates on the lab bench in front of you (make sure they are right-side up). You should have a lighted burner within reach.

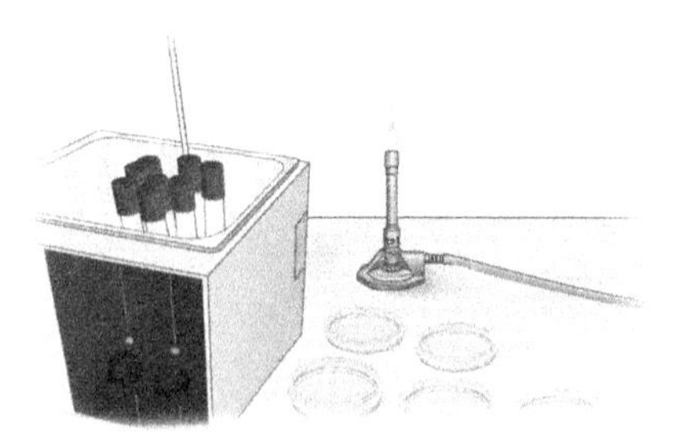

4. Remove a test tube with melted agar from the water bath. Wipe excess water from the outside of the tube with a paper towel.

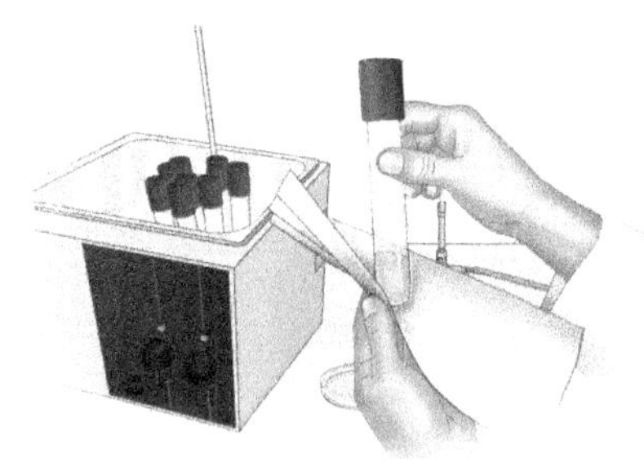

5. Hold the tube at an angle and remove the cap. Quickly pass the open mouth of the tube three times through the flame.

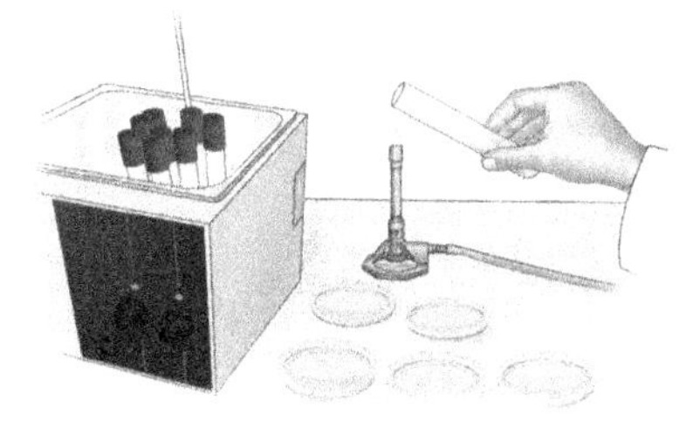

6. Lift the plate lid, just enough to pour—quickly—all of the melted agar in the tube into the plate.

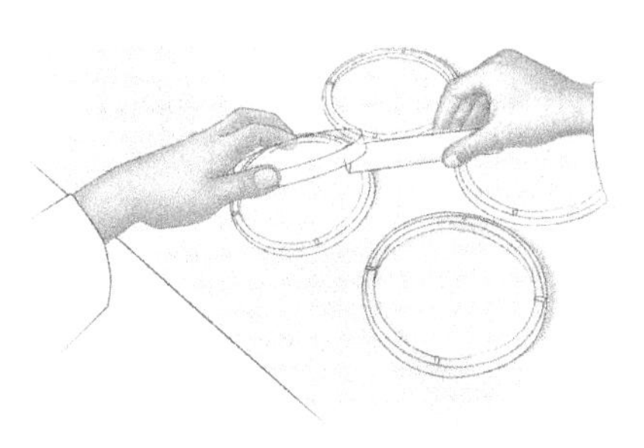

7. To ensure that the melted agar covers the bottom of the plate, gently rotate the plate in a small circle on the bench top.

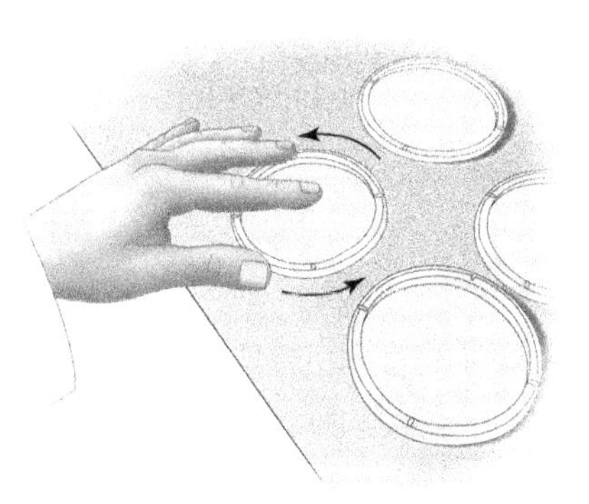

8. After the agar has cooled in the plates, leave the plates out overnight if you will not use them immediately. This reduces condensation during storage. Store them upside-down in a bag to prevent the agar from drying out in the refrigerator.

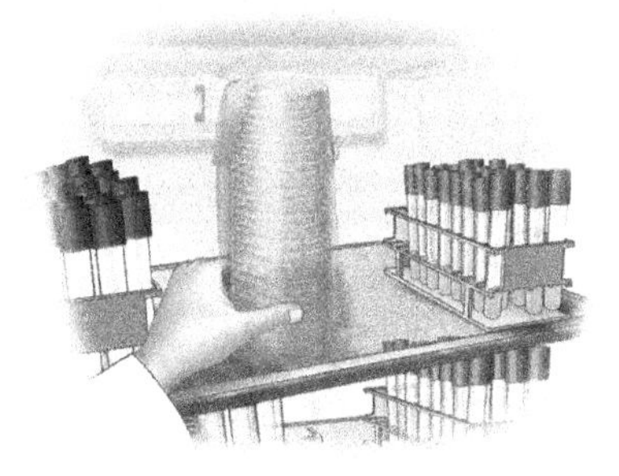

Method II: Pouring sterile melted agar from a flask into plates

1. After you have weighed out the correct amount of dehydrated medium, add the powder to a flask with the measured amount of distilled or deionized water. The flask should hold at least twice the volume of water used. If the total volume is >1 L, prepare media in 500 mL aliquots.

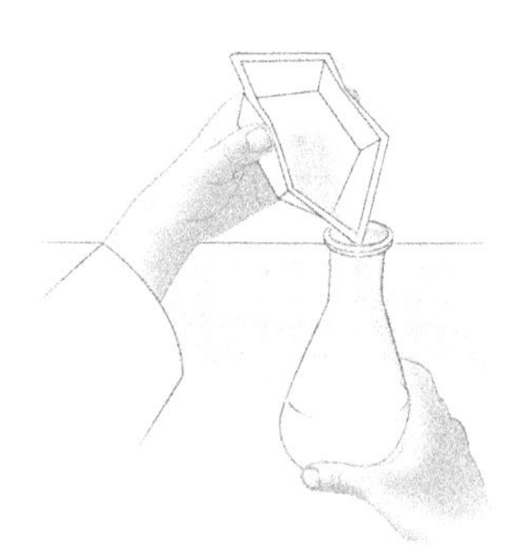

2. Swirl the contents of the flask to disperse the powdered medium in the water. Loosely cover the flask mouth with a large aluminum foil square. Sterilize the agar medium in the autoclave. The agar will melt at the high temperature in the autoclave.

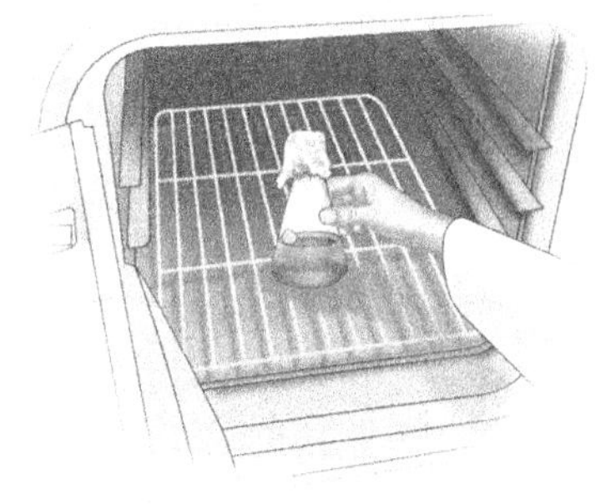

3. When the sterilization cycle is complete in the autoclave, allow the flask to remain in the chamber for at least 10 minutes. You must wear a protective glove when you remove the flask.

4. Cool the agar in a 48–50° C water bath. If a water bath is not available, cool until the flask is hot to the touch but does not burn your fingers. While the agar is cooling, arrange the Petri plates on the bench top.

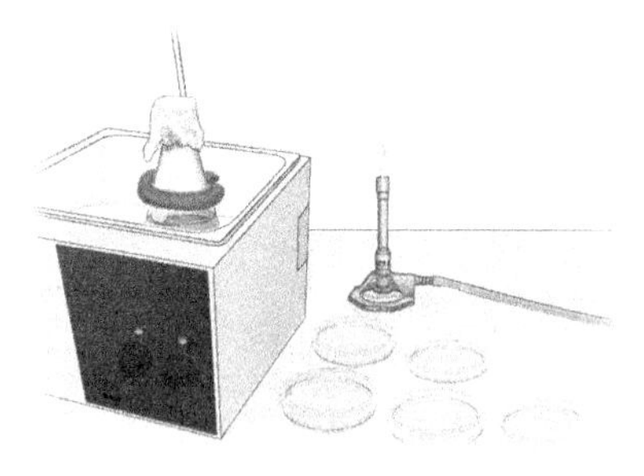

5. When the agar has cooled but is still liquid, remove the foil cover from the flask and pass the mouth through a flame.

6. Lift each plate lid and pour enough melted medium to cover most of the plate's bottom.

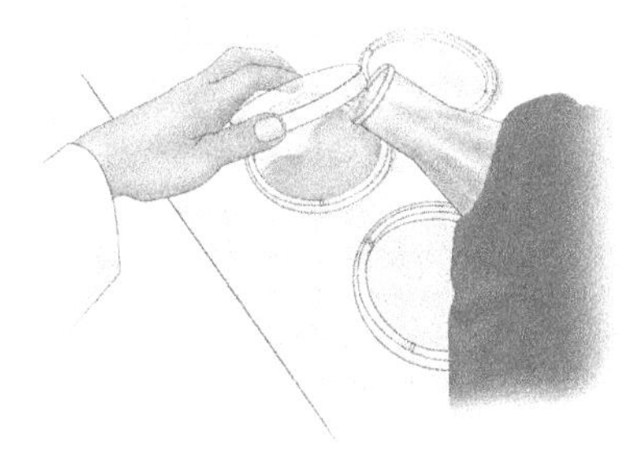

7. To ensure that the melted agar covers the bottom of the plate, gently rotate the plate in a small circle on the bench top.

8. If bubbles appear in the melted agar, you can eliminate them by a brief exposure to the flame from a burner.

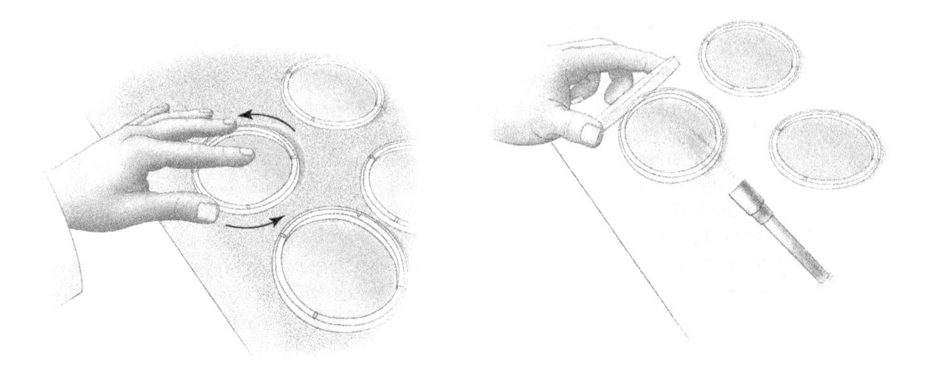

9. After the agar has cooled, leave the plates out overnight if you will not use them immediately. This reduces condensation during storage. Store them upside-down in a bag to prevent the agar from drying out in the refrigerator.

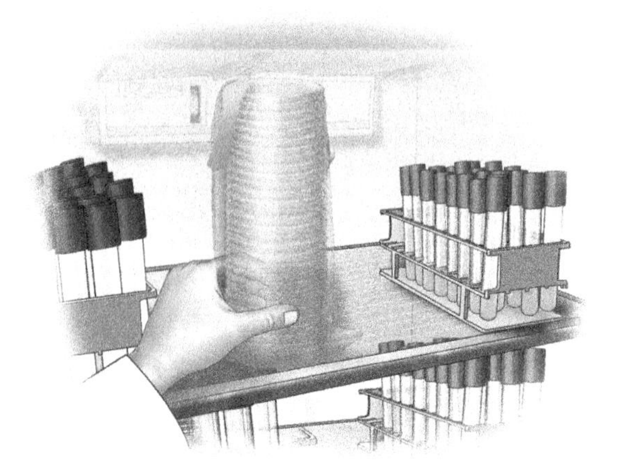

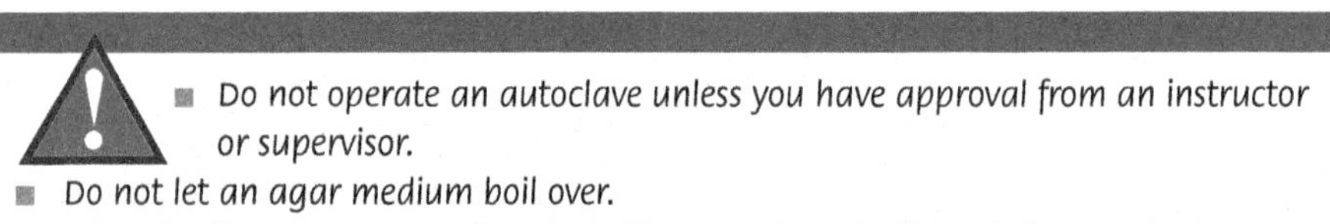

- Do not operate an autoclave unless you have approval from an instructor or supervisor.
- Do not let an agar medium boil over.
- When heating an agar medium to boiling, or when sterilizing it in an autoclave, use a flask or beaker that has a volume twice that of the volume of the medium.
- Wear insulated or heat-proof gloves when you unload hot materials from an autoclave or handle glassware containing hot agar.
- When you are using an open flame, position it away from your hair and clothing.
- If you are burned, seek immediate treatment.

Sterilization

Purpose:

- To destroy all microorganisms that may contaminate materials used in the microbiology laboratory.

Three procedures are routinely used in the microbiology lab to sterilize media and materials that must be free of contaminating organisms. They include the autoclave, dry heat, and filtration.

The Autoclave

An autoclave creates moist heat at temperatures high enough to kill bacterial endospores. Steam under pressure at 15 lb/in^2 creates the needed temperature of 121° C. Materials sterilized in an autoclave must be able to withstand the high temperature without breaking down or melting. Media, glassware, and cotton swabs are routinely sterilized in an autoclave; it is also used to sterilize contaminated glass tubes and Petri plates. Plastic Petri plates are sterilized in autoclaveable plastic bags. The sterilization time for such biohazard waste is at least 45 minutes.

Some labs are equipped with an autoclave with electronic controls and a self-contained steam generator. Time and temperature are pre-programmed. Other labs may have an autoclave that requires filling a reservoir with deionized or distilled water and has dials for setting the time and temperature.

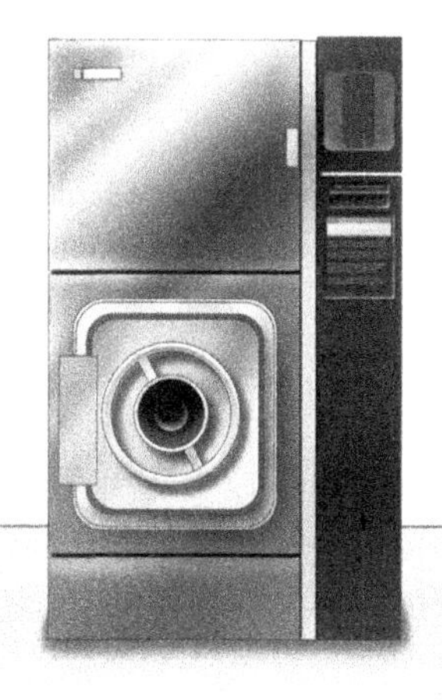

Autoclave

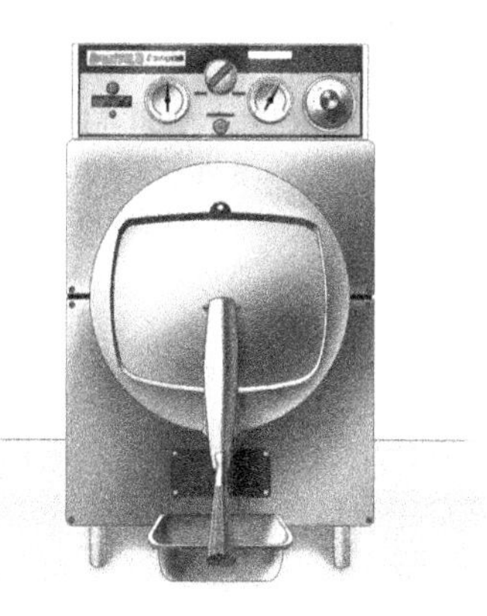

Autoclave with reservoir

Guidelines for successful sterilization with an autoclave

- Paper wrap is preferred over aluminum foil wrap for sterilizing dry objects because steam cannot penetrate foil.
- Larger volumes of liquids or packs of dry materials require a sterilization time longer than 20 minutes to allow the steam time to penetrate.
- A liquid medium must be autoclaved in a container that has twice the volume of the liquid.
- If the autoclave is operated manually, air must be expelled from the chamber to ensure that only steam pressure is measured.
- Substances that cannot be penetrated by steam, such as paraffin, mineral oil, or petroleum jelly, cannot be sterilized in an autoclave; dry heat is the preferred method.
- Special tape attached to wrapping or baskets will indicate whether materials have been successfully sterilized with a color change or the appearance of words (e.g., "sterile").
- The door of the autoclave must be tightly sealed.
- In a manually operated autoclave, choose the "slow exhaust" option for liquids; choose "fast exhaust" for dry materials.

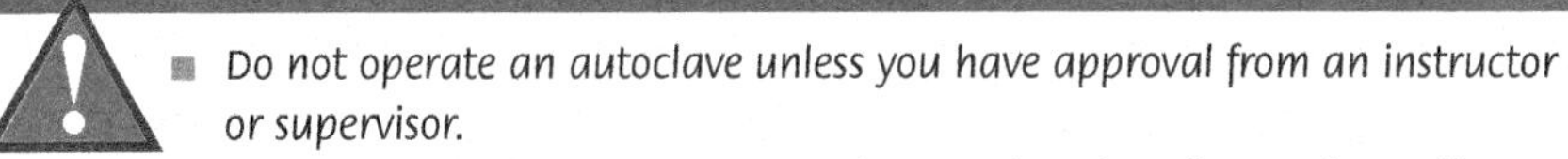

- *Do not operate an autoclave unless you have approval from an instructor or supervisor.*
- *Only borosilicate or soda lime glassware can be autoclaved: ordinary glass will not withstand the high temperature.*
- *To avoid scalding your face and arms, open the door slowly, only an inch, at the end of a sterilizing cycle.*
- *Wait 10 minutes before removing objects from an autoclave. Superheated water will boil after autoclaving if the liquid is even slightly agitated, and it could scald you.*
- *Wear insulated or heat-proof gloves when you unload hot materials from an autoclave.*
- *If you are burned, seek immediate treatment.*

Dry Heat

Dry heat is preferred for sterilizing glassware, metal objects, and substances such as petroleum jelly, mineral oil, and paraffin. A hot air oven reaches the required temperature of 160–180° C (320–356° F). Materials are heated for 2–3 hours. Note that cotton and paper will char at temperatures higher than 180° C.

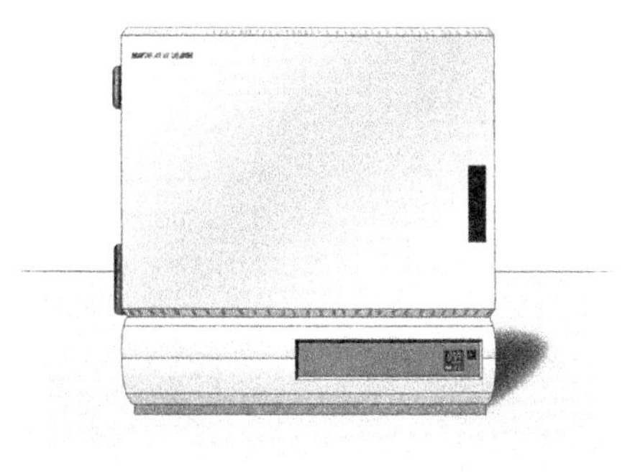

Hot air oven

■ *Wait at least an hour after a dry heat sterilization cycle is completed before you remove glassware or other materials from the oven. If possible, sterilize with dry heat overnight using a timer that turns the oven on and off at appropriate times. Glassware will be cool when you return to the lab in the morning.*

Filtration

Filtration sterilizes by physically removing microorganisms from solutions with solutes that cannot withstand high temperatures. Heat-sensitive solutes include antibiotics, certain sugars, enzymes, and growth factors, such as vitamins.

Filters are made of a synthetic polymer or a cellulose ester formed into a microscopic mesh. The pores in this mesh (typically 0.22 µm [preferred] or 0.45 µm) allow passage of a solution but block passage of contaminating bacteria and fungi. Viruses are too small to be blocked with the filters usually used in microbiology.

A disposable syringe filter can be used to sterilize small volumes (1–20 mL). Fluid added to the syringe barrel is forced through the filter by pushing the plunger. The sterile liquid is collected in a sterile container.

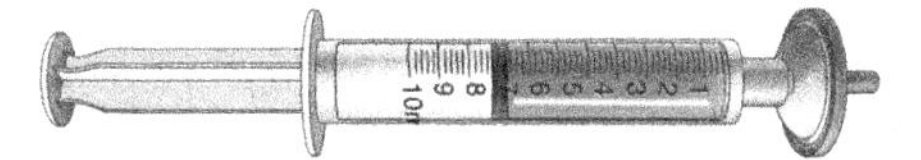

Syringe filter

Procedure:

1. Before you connect the filter to the syringe, remove the plunger. After you have fastened the syringe tip onto the filter, pour the nonsterile liquid into the syringe barrel.

2. Insert the plunger into the barrel. Push the plunger to force the liquid through the filter and into a sterile container.

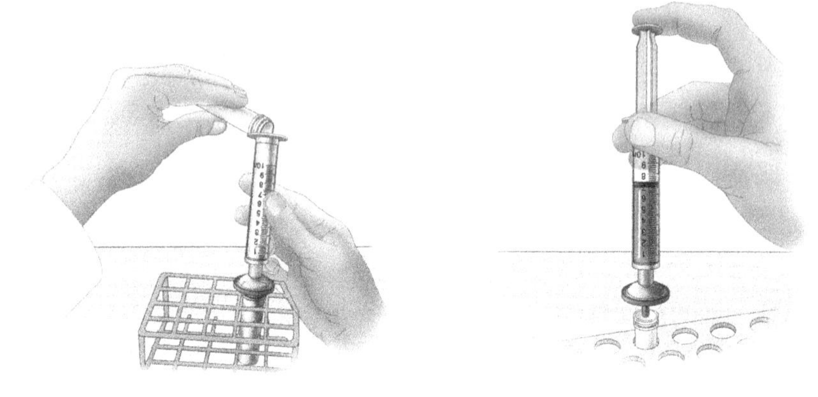

A disposable membrane filtration unit is appropriate for sterilizing larger volumes (25 mL to 1 L). A vacuum pulls the nonsterile liquid through the filter into a sterile container. A disposable unit is convenient but can be expensive.

Procedure:

1. Secure the filtration unit with a clamp. Connect the unit to a vacuum pump.
2. Add the nonsterile liquid to the top of the unit. Turn on the vacuum pump.

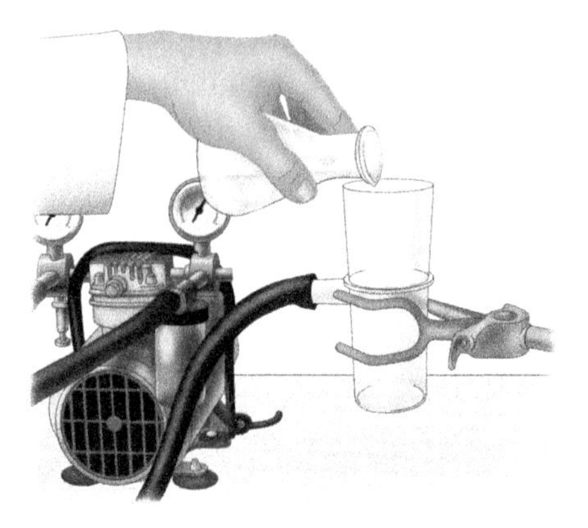

Disposable membrane filter unit

A reusable membrane filter apparatus is also available. The membrane and the reusable glass components are individually sterilized in the autoclave and stored until needed. Aseptic technique is used to add the membrane and assemble the components.

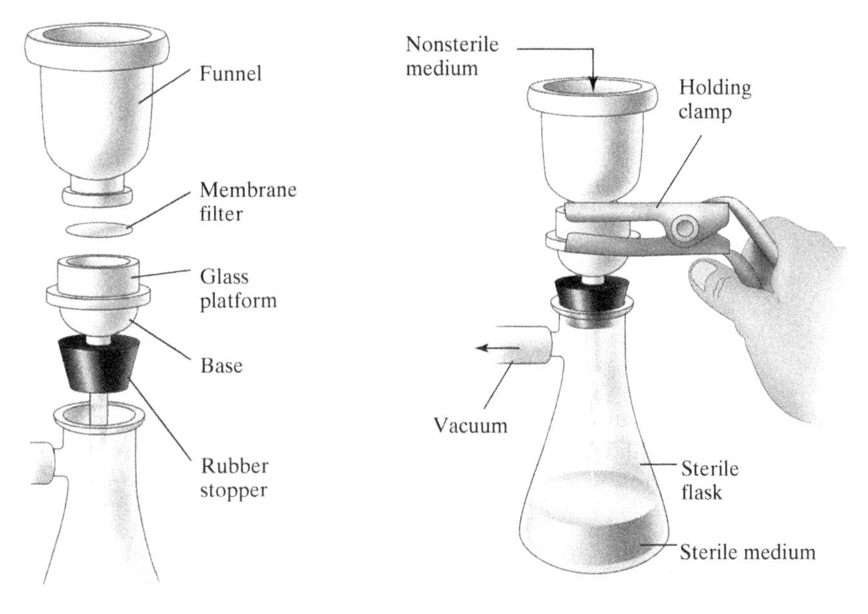

Reuseable membrane filter unit

Procedure:

1. Use a sterile forceps to remove a sterile filter membrane from its packet. Place the filter in the center of the glass platform already in place on the sterile flask.

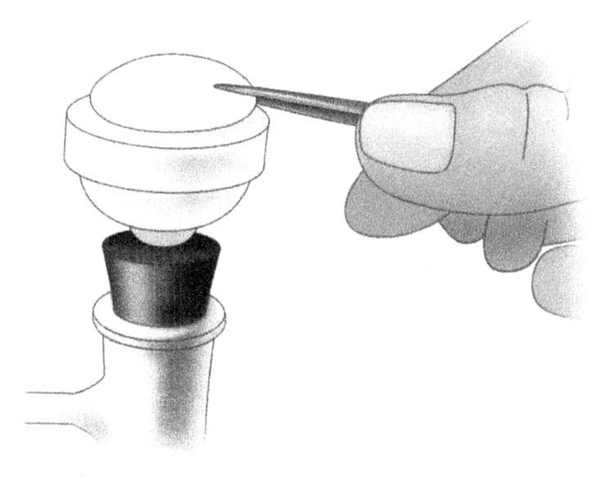

2. Use a clamp to secure the top funnel to the glass platform.

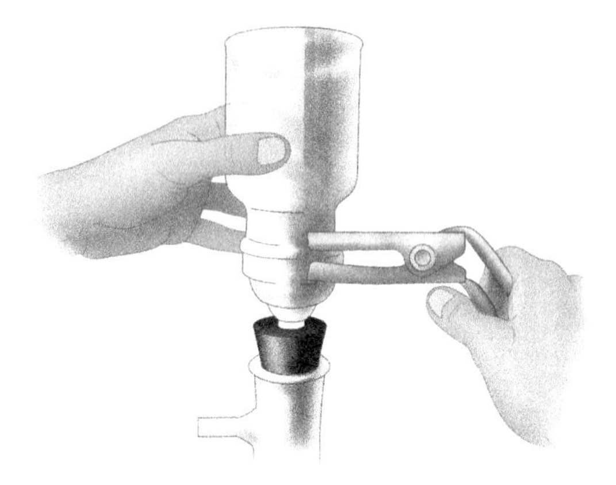

3. Pour a small amount of nonsterile solution in the top funnel to moisten the membrane filter. Turn on the vacuum and slowly add the remaining nonsterile solution to the top funnel. Turn off the vacuum when all of the solution has passed through the filter.

Adjusting the Gas Burner

Purpose:

- To develop effective skills in lighting a gas burner and generating an optimal flame.

Materials:

- gas source (methane [natural gas] or propane [bottled gas])
- Bunsen burner or Fisher (grid) burner
- flint striker (sparker) or matches

The optimal flame for sterilizing inoculating loops and needles, for fixing bacterial smears, and for flaming the opening of tubes and other glassware looks like the figure below. It includes a bright pale blue cone (about 1 $^1/_2$ in. or 4 cm) that is surrounded by a translucent blue/violet bushy outer flame (about 2 $^1/_2$ in. or 6 cm). The hottest spot is found just above the top of the inner cone.

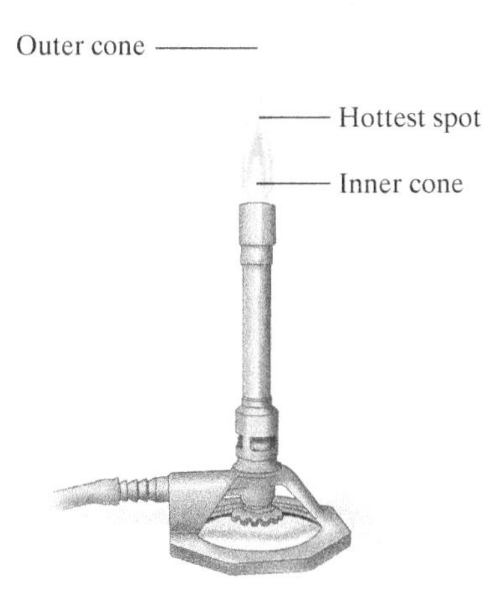

Features of an optimal gas flame

The Bunsen burner has two control points for creating an optimal flame. The toothed wheel in the base is rotated to adjust gas flow into the burner. Turning this wheel clockwise increases gas flow. Counterclockwise decreases gas flow. The barrel, or tube, is turned to regulate the size of the openings in the air holes. Less air enters the tube with clockwise turns and more air with counterclockwise turns. Both of these control points must be adjusted in order to create the most desirable air/gas mixture.

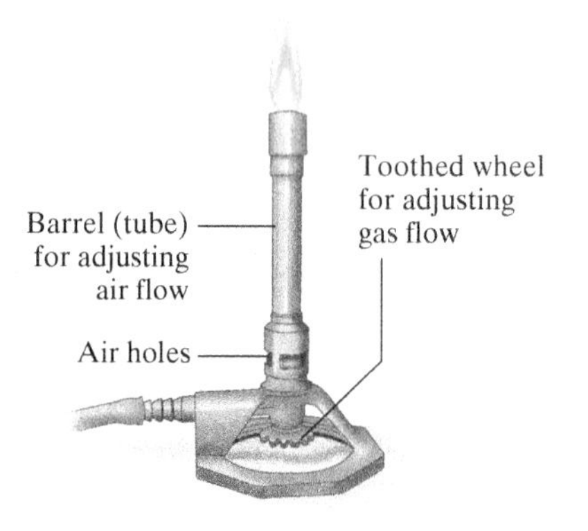

Features of the Bunsen burner

Procedure:

1. Connect the rubber tubing on the burner to the gas outlet. Turn on the gas by positioning the handle parallel to the outlet nozzle.

2. Turn the toothed wheel a quarter turn in each direction to ensure that gas is flowing. You should hear a quiet hissing from the opening.

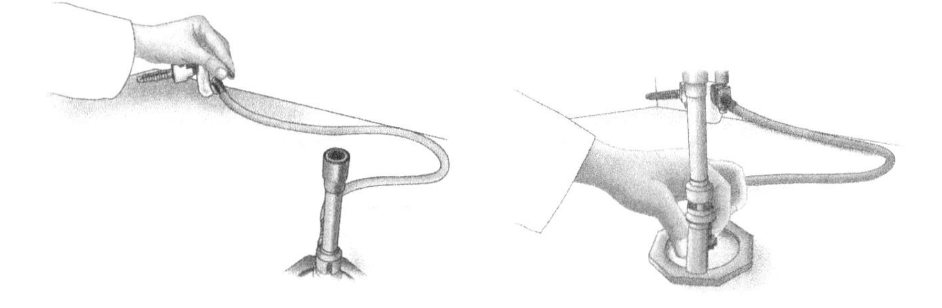

3. Hold the flint striker just above and to the side of the tube opening. With the striker at a 45° angle, squeeze vigorously enough to raise sparks. Do this until the gas ignites.

3a. If you are using a match to light the burner, hold the lit match just above and to the side of the barrel opening.

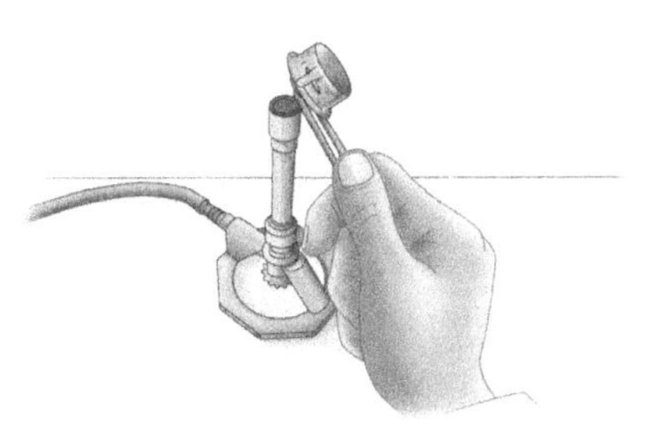

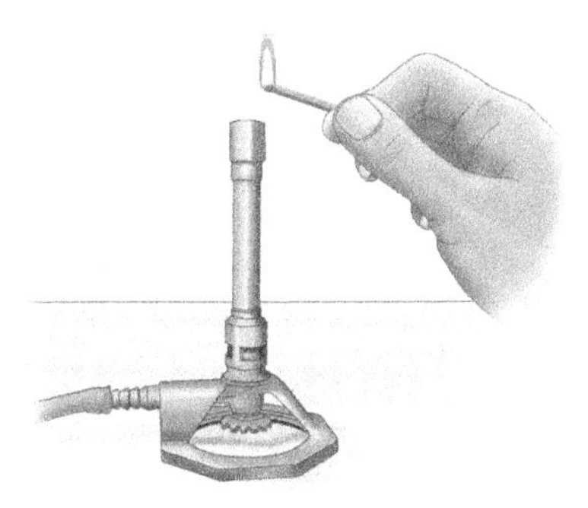

4. If the burner does not light after a few attempts, turn off the gas and turn the toothed wheel in the base a quarter to a half turn clockwise. Turn on the gas again and repeat Step 3.

5. To get an optimal flame after the burner is lit, adjust the airflow by twisting the barrel and adjust the gas flow by turning the toothed wheel.

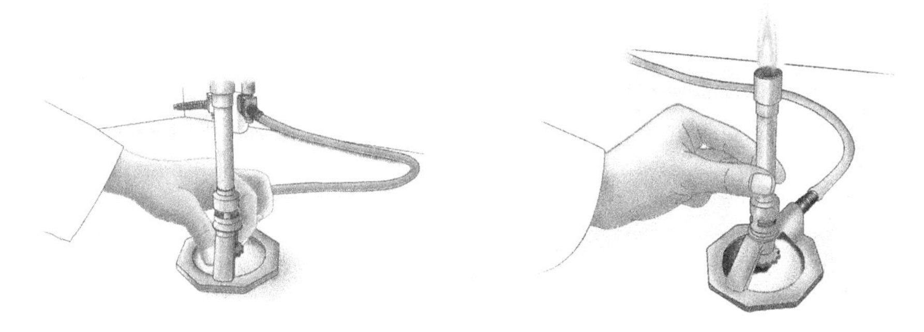

6. The Fisher (grid) burner is adjusted in a similar fashion to get a desirable flame.

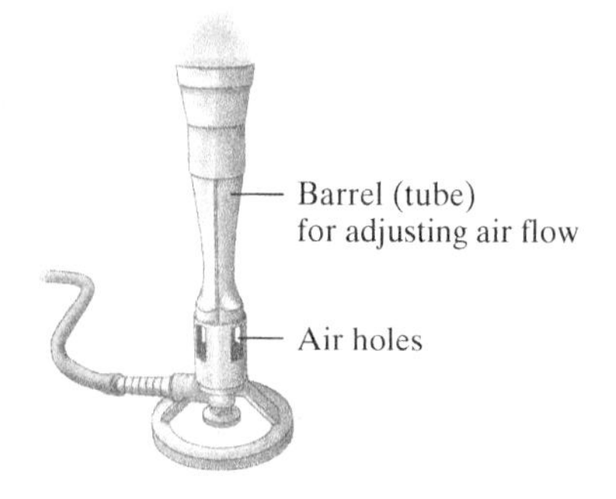

Troubleshooting

- If the handle at the gas outlet is not turned so that it is parallel to the nozzle, gas flow from the source will not be sufficient.
- If sparks do not appear while you squeeze the striker:
  - squeeze the handle more tightly.
  - check the flint. If it has worn away, replace the flint.
- If the inner cone is too large or is not discernable, too much air is flowing into the burner. Reduce airflow by twisting the barrel clockwise.
- If the flame appears yellow or orange, too little air is flowing into the burner. Increase airflow by twisting the barrel counterclockwise
- If the flame is not high enough, increase gas flow by turning the toothed wheel clockwise.
- If the flame is too high, decrease gas flow by turning the toothed wheel counterclockwise.

- *Wear safety goggles or other eye protection when you light the gas burner and when you use the flame.*
- *Before you light the burner, check the tubing for cracks and tears. Replace damaged tubing to prevent gas leakage.*
- *Position the burner so it is accessible but so you will not burn yourself or your clothing.*
- *Tie back long hair to prevent it catching on fire when you are using an open flame.*
- *Turn off the gas when a flame is not needed.*
- *When you turn off the gas, make sure that the handle on the gas outlet is perpendicular to the outlet nozzle.*
- *If you use a match to ignite the gas, make sure you run water on the match before you dispose of it.*

Using an Electrical Incinerator (Bacti-Cinerator)

An electrical incinerator can be used to sterilize loops, wires, and the mouths of test tubes, much like a flame.

Loops and wires: Hold the loop or wire inside the incinerator for 5–7 seconds.

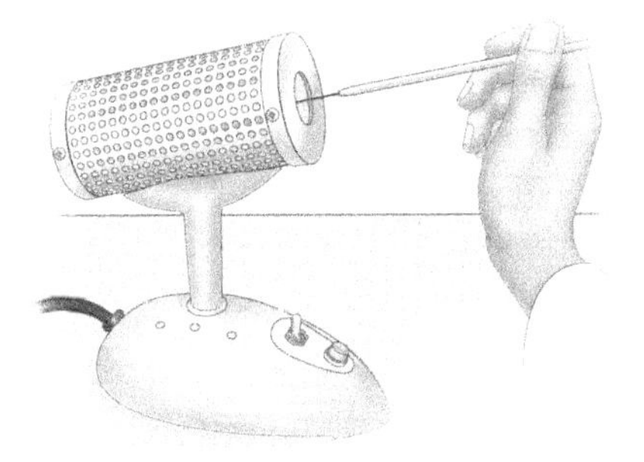

Culture tube mouth: Use the heat at the opening of the incinerator to sterilize a culture tube mouth. Hold the tube mouth next to incinerator opening for 5–7 seconds.

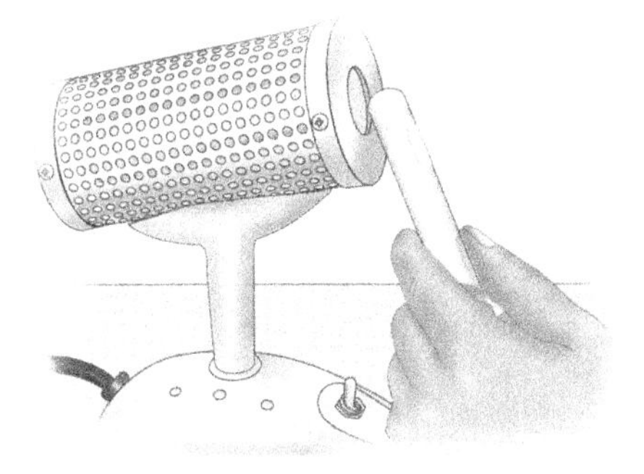

Aseptic Transfer of Bacteria

Preparing Bacterial Cultures in Broth Tubes and on Agar Slants

Purpose:

- To inoculate culture media with specific bacteria without introducing contaminating microbes. This is known as **aseptic technique**.

Materials:

- broth culture of bacteria
- sterile nutrient broth tube
- sterile nutrient agar slant
- sterile nutrient agar plate
- inoculating loops and needles

Inoculating loop

Inoculating needle

Procedure:

Preparation of the Work Area

1. Disinfect the entire work area by spreading disinfectant (5% Amphyl, or quaternary ammonium compound diluted per manufacturer's instructions). Allow the surface to air dry: do not wipe dry.

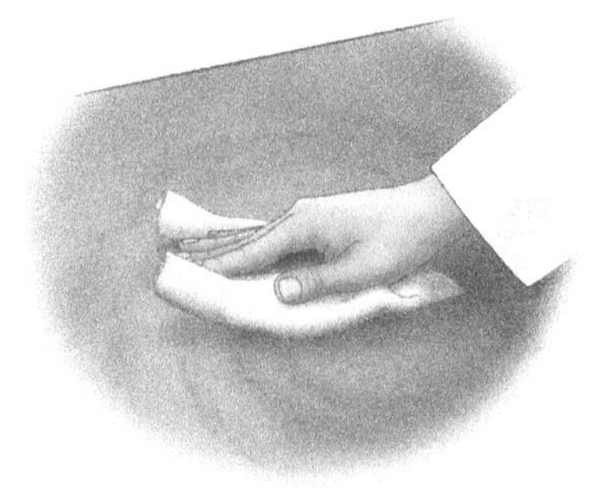

2. Arrange materials for efficiency and safety. Place the burner to minimize risk of burning clothing or hair. Set the test tube rack on the opposite side of the workspace. Keep books and papers out of the workspace.

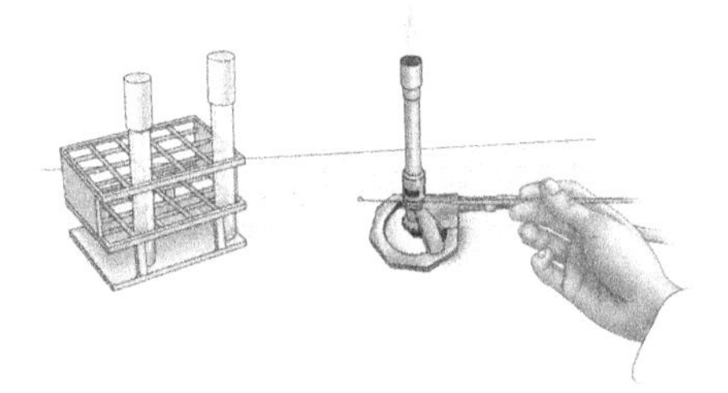

Transferring from a Broth Culture into a Broth Tube

1. Label a sterile broth tube with the organism's name, the date, and your name or initials.

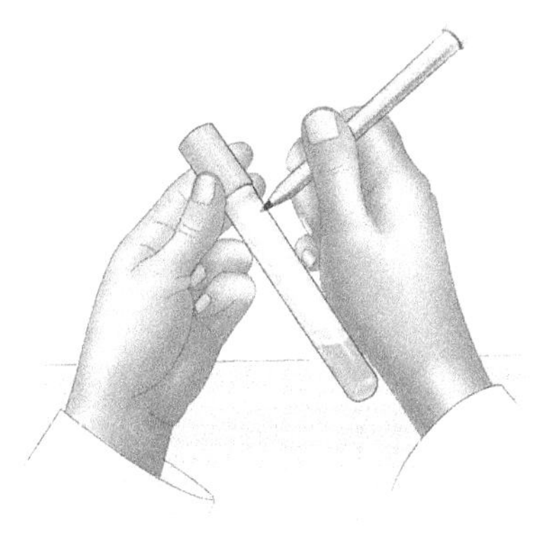

2. Suspend bacteria in the culture tube by gently shaking the tube. Do not hold the tube by the cap: it might come off.

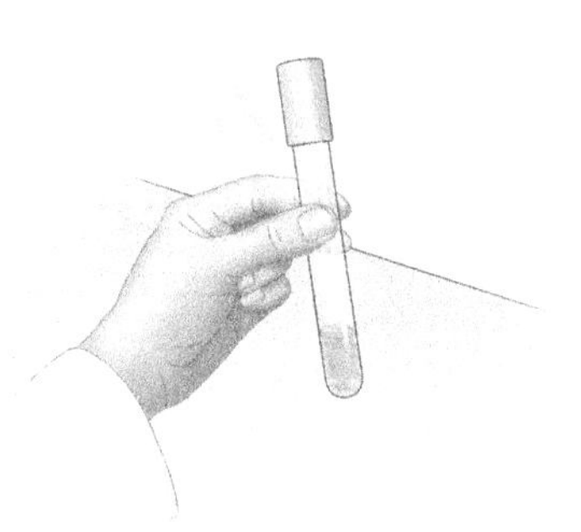

3. Hold the loop as you would a pencil. Sterilize the loop by flaming the wire until it is red-hot along its length, not just the loop. Allow the loop to cool but do not let it touch anything.

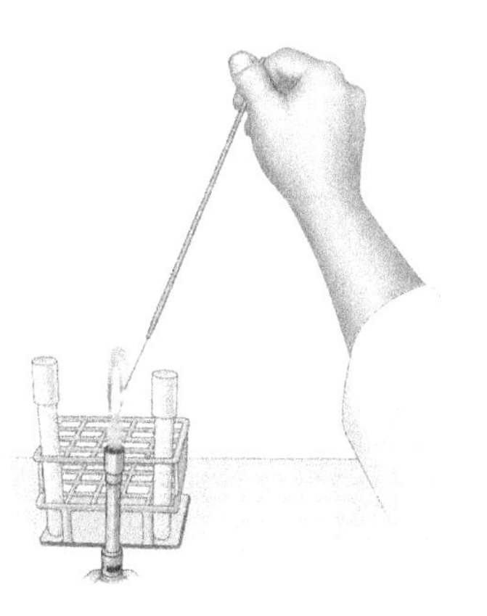

4. Hold the culture tube in your other hand. Curl the little finger of the hand holding the inoculating loop around the cap and remove it by twisting the tube with your other hand. Do not set the cap down.

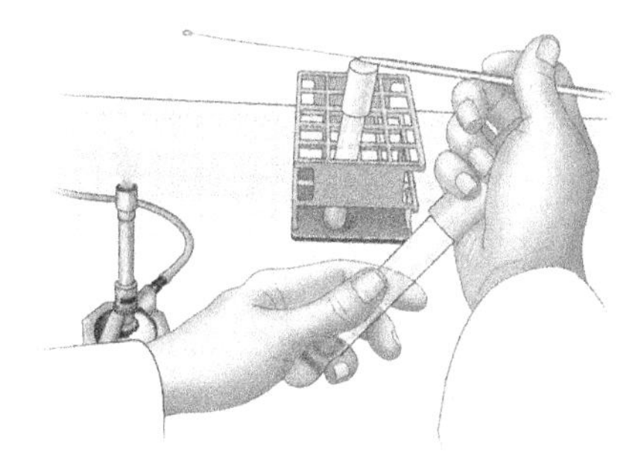

5. Quickly pass the tube mouth through the flame three times. Hold the tube at an angle to minimize the chance of dust particles falling into the open tube.

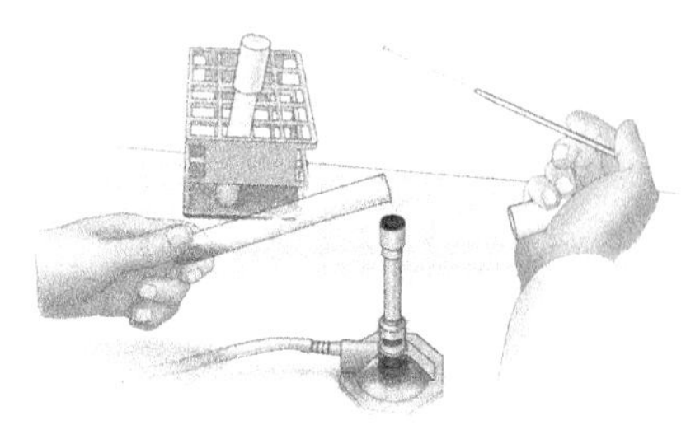

6. Insert the cooled loop into the broth culture and withdraw a bead of culture held within the loop.

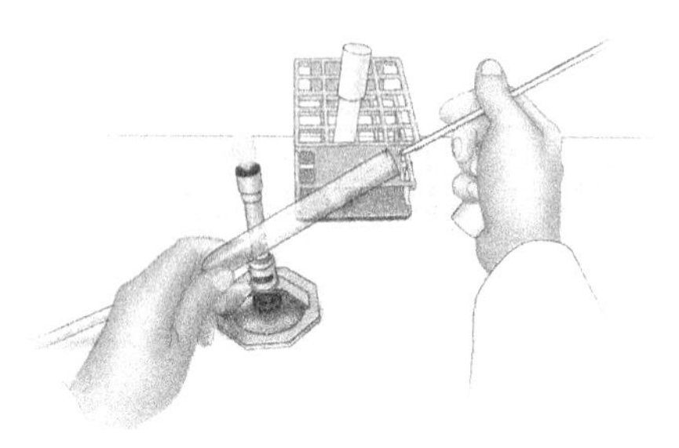

7. Flame the tube mouth. Replace the cap on the culture tube and set it in the test tube rack. Take care that you do not dislodge the bead of culture in the loop.

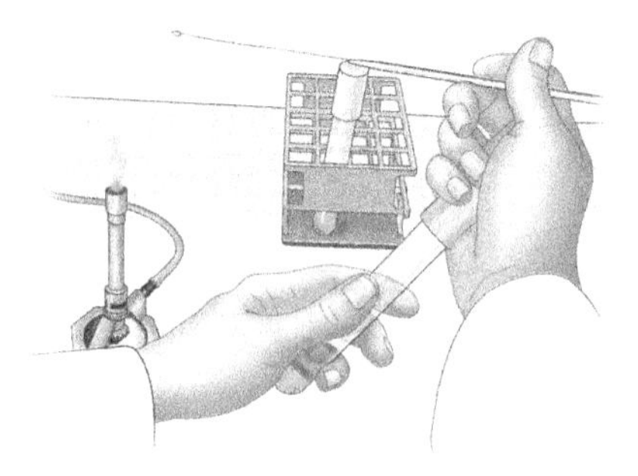

8. Pick up the labeled sterile broth tube and remove the cap as you did for the other tube. Pass the tube opening through the flame.

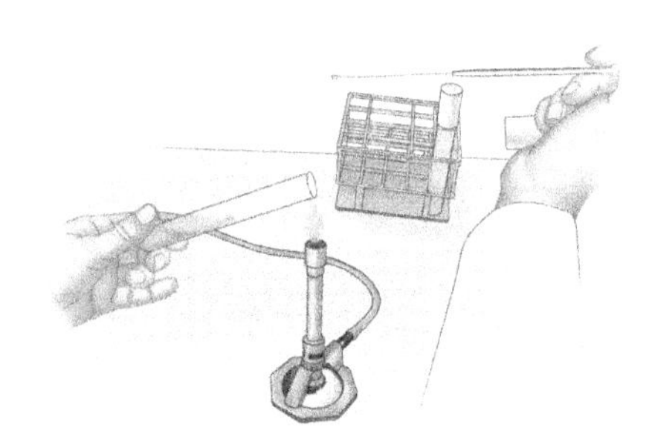

9. Insert the loop with bacteria into the fresh broth and roll the loop handle between your fingers to dislodge the bacteria.

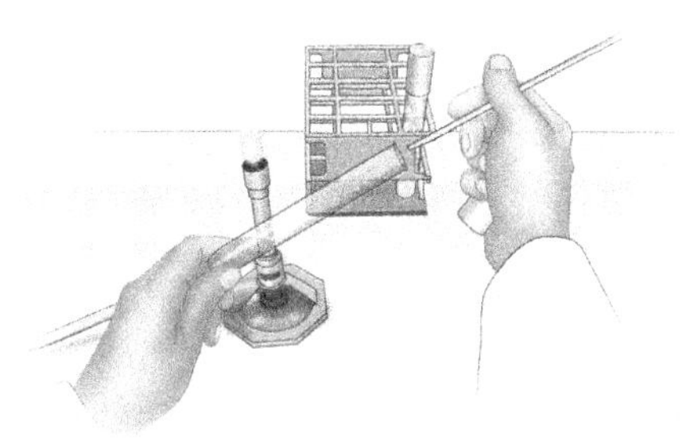

10. Flame the broth tube mouth. Replace the cap on the tube and set it in the test tube rack.

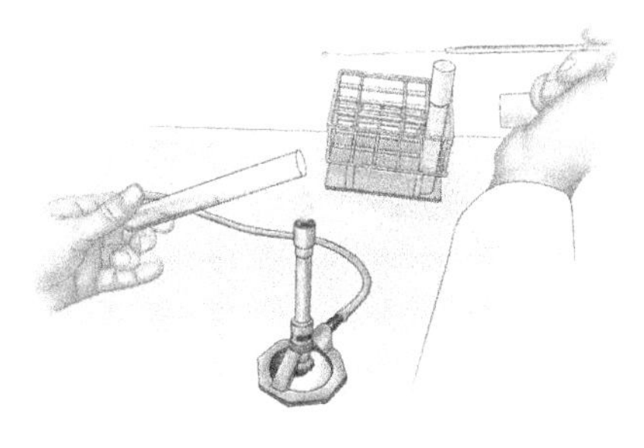

11. Flame the loop and wire until red-hot. Set it down to cool.

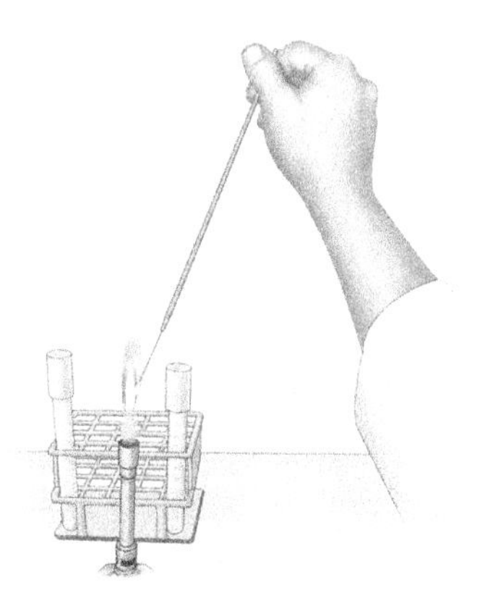

12. Place the newly inoculated tube in the incubator.

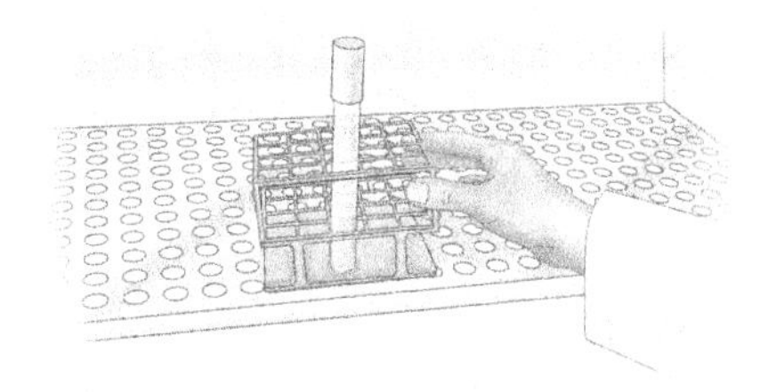

Transferring from a Broth Culture onto an Agar Slant

1. Label a sterile agar slant tube with the organism's name, the date, and your name or initials.

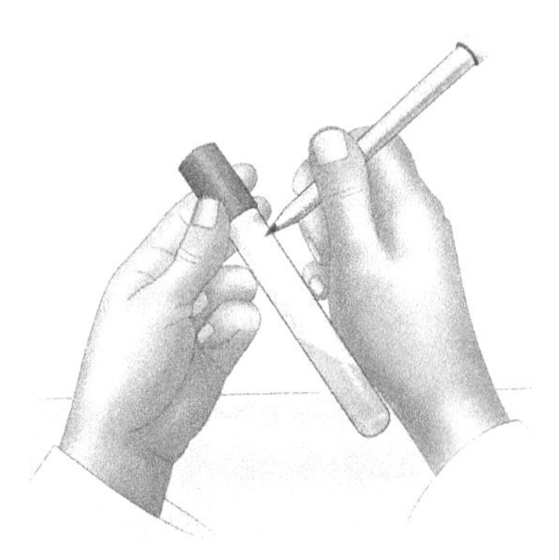

2. Suspend bacteria in the culture tube by gently shaking the tube. Do not hold the tube by the cap: it might come off.

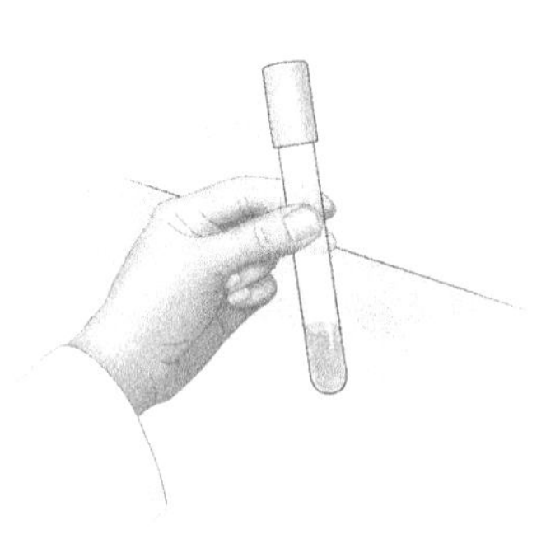

3. Hold the loop as you would a pencil. Sterilize the loop by flaming the wire until it is red-hot along its length, not just the loop. Allow the loop to cool but do not let it touch anything.

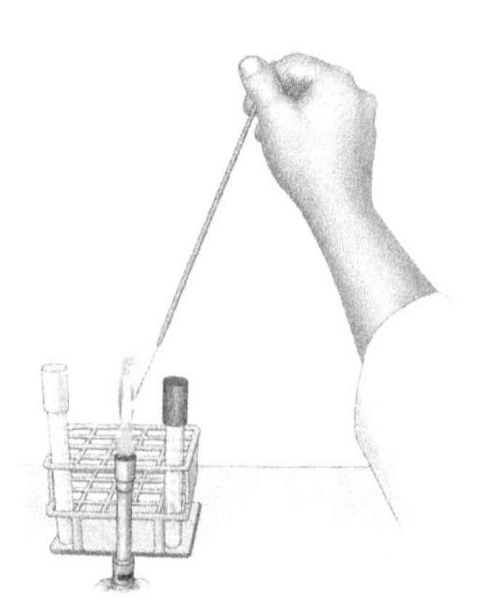

4. Hold the culture tube in your other hand. Curl the little finger of the hand holding the inoculating loop around the cap and remove it by twisting the tube with your other hand. Do not set the cap down.

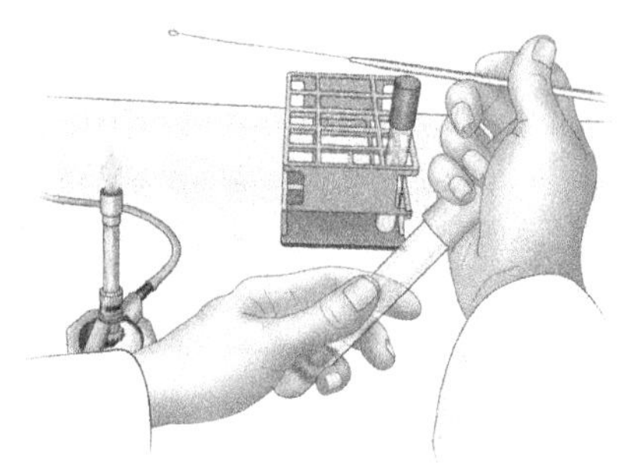

5. Quickly flame the tube mouth. Hold the tube at an angle to minimize the chance of dust particles falling into the open tube.

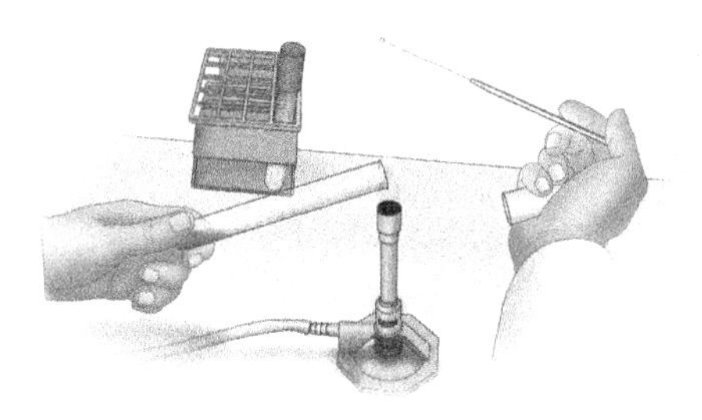

6. Insert the cooled loop into the broth culture and withdraw a bead of culture held within the loop.

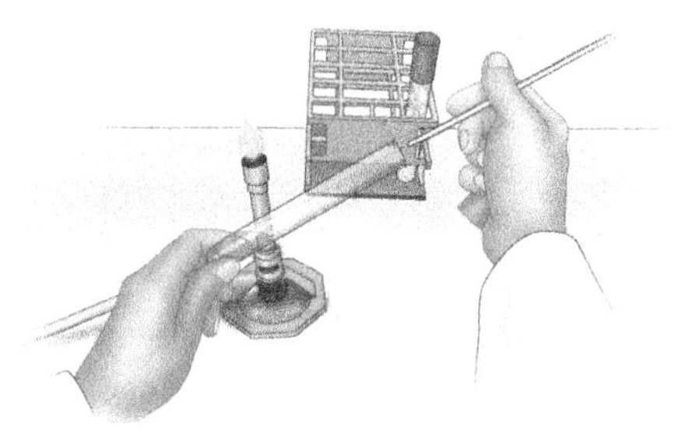

7. Flame the tube mouth. Replace the cap on the culture tube and set it in the test tube rack. Take care that you do not dislodge the bead of culture held in the loop.

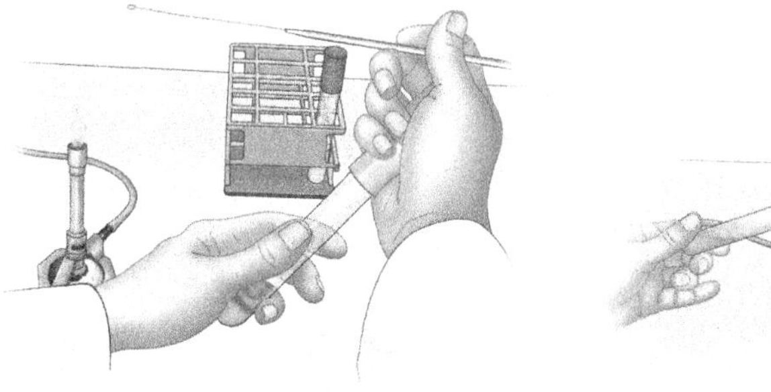

8. Pick up the labeled sterile agar slant tube and remove the cap as you did for the culture tube. Pass the tube opening through the flame.

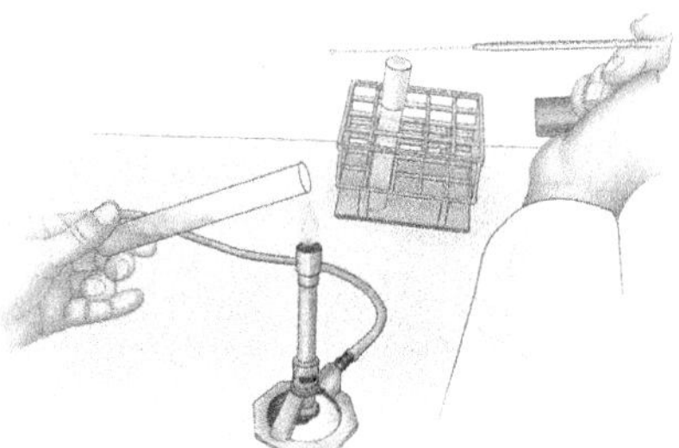

9. Inoculate the slant by gently moving the loop back and forth over the agar surface from the bottom to the top.

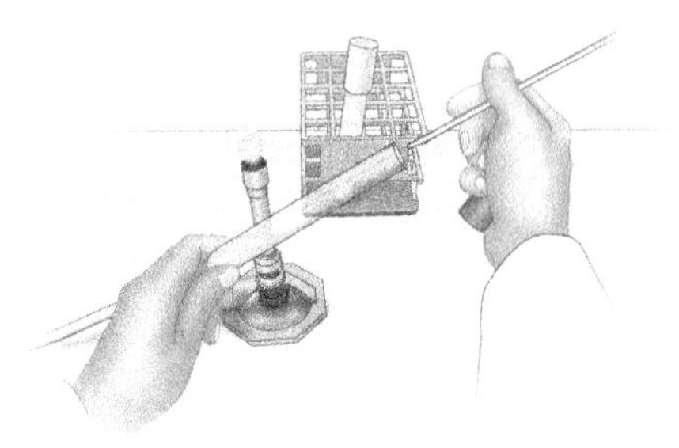

10. Flame the slant tube mouth. Replace the cap on the tube and set it in the test tube rack.

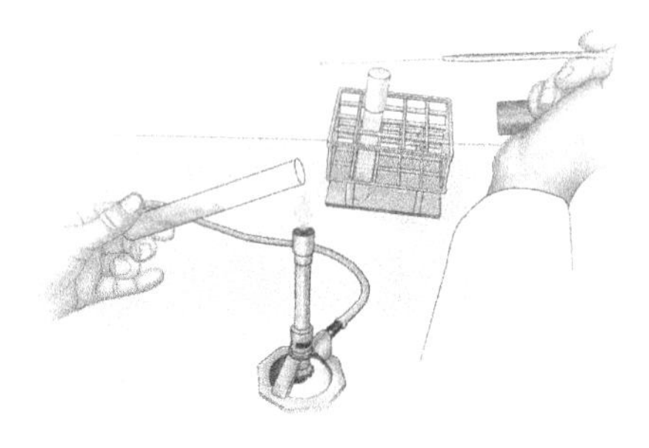

11. Flame the loop and wire until red-hot. Set it down to cool.

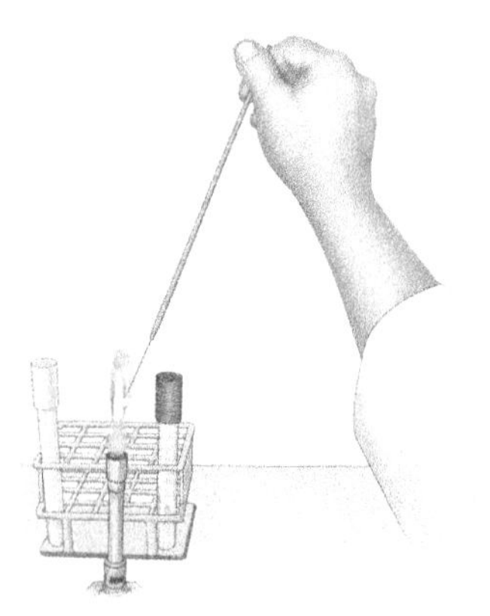

12. Incubate the newly inoculated tube in the incubator.

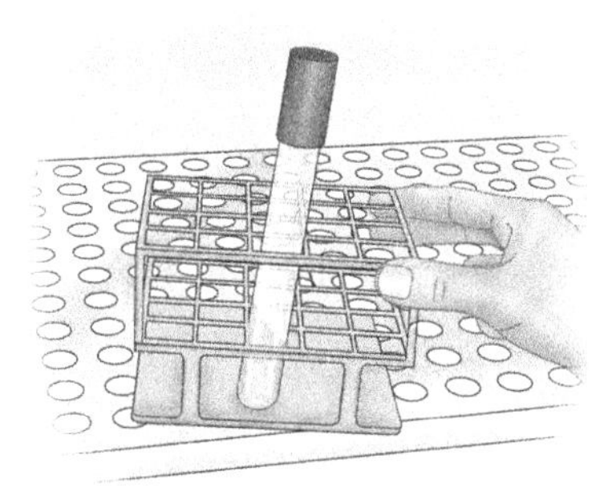

Transferring a Bacterial Colony from an Agar Plate to an Agar Slant

1. Select an isolated colony on an agar plate surface and write an identifying number under it on the bottom of the plate.

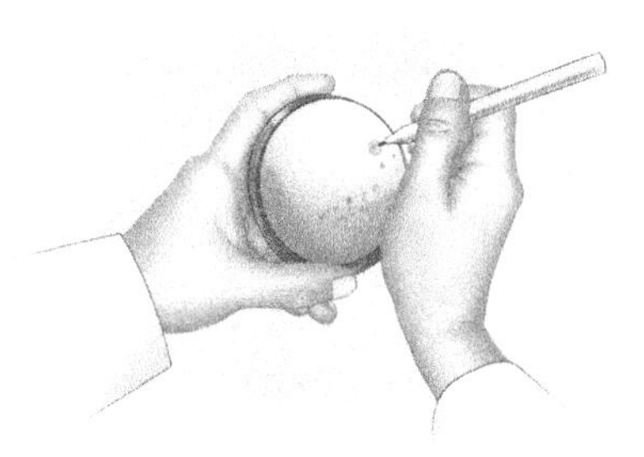

2. Label an agar slant with that identification number, the date, and your name or initials. Place it nearby in a test tube rack.

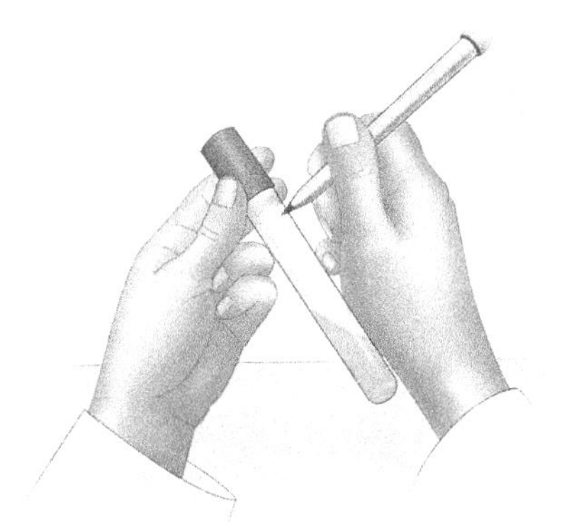

3. Sterilize the loop by flaming the wire until it is red-hot along its length, not just the loop. Allow the loop to cool but do not let it touch anything.

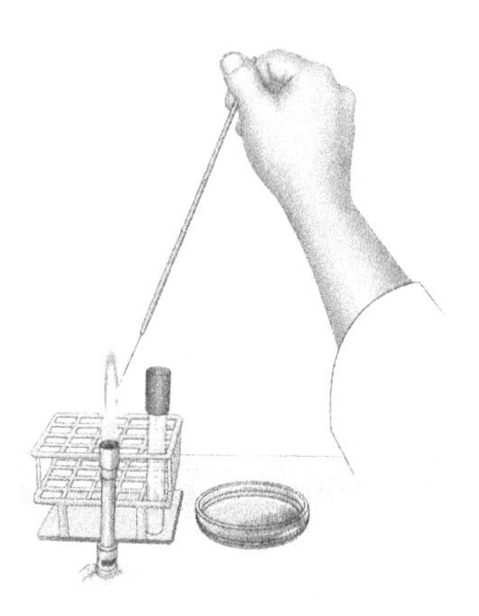

4. Lift the Petri plate lid just enough to pick up the identified colony on the loop.

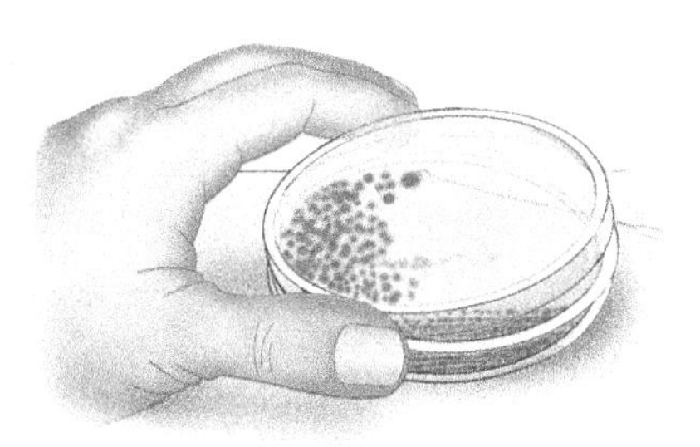

5. While you are holding the inoculating loop handle, pick up the labeled slant tube. Remove the cap and pass the tube opening through the flame.

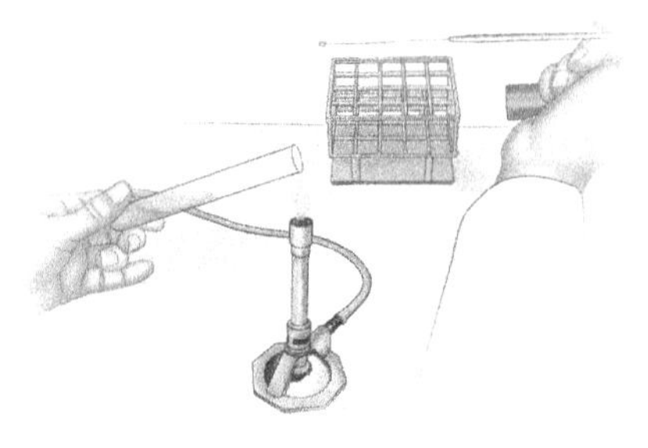

6. Inoculate the slant by gently moving the loop back and forth over the agar surface from the bottom to the top.

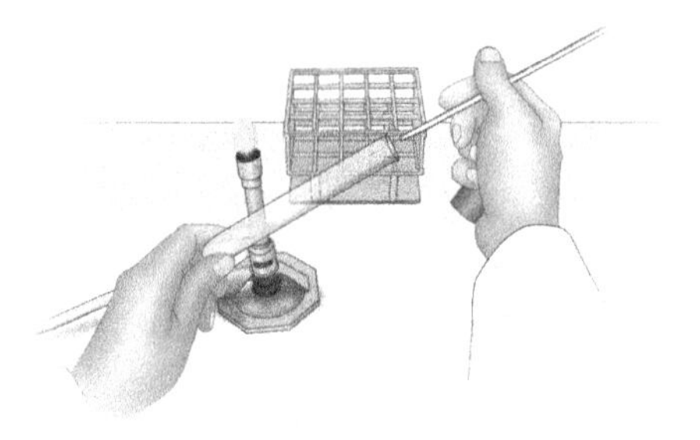

7. Flame the slant tube mouth. Replace the cap on the tube and set it in the test tube rack.

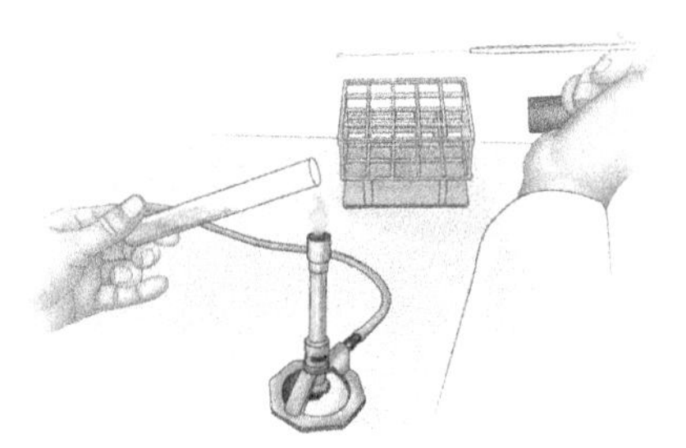

8. Flame the loop and wire until red-hot. Set it down to cool.

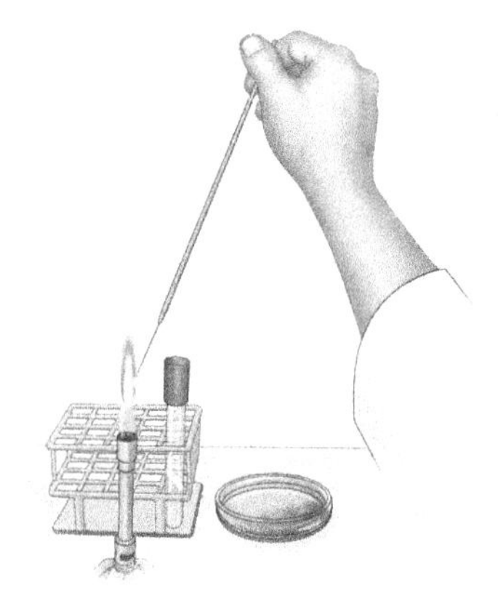

9. Incubate the newly inoculated tube in the incubator.

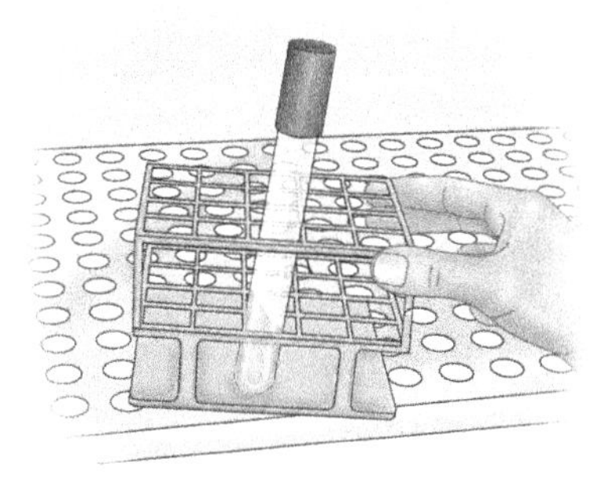

Transferring Bacteria while Holding Multiple Tubes

1. Hold both the original culture tube and the tube with fresh medium in one hand. Place the tube with the culture farther from you.

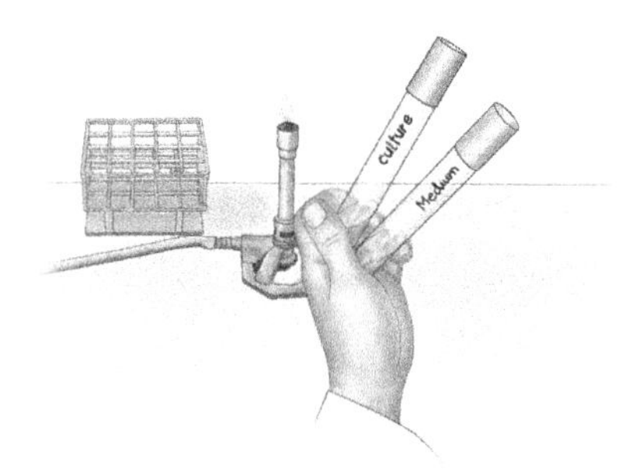

2. Flame the inoculating loop or needle.

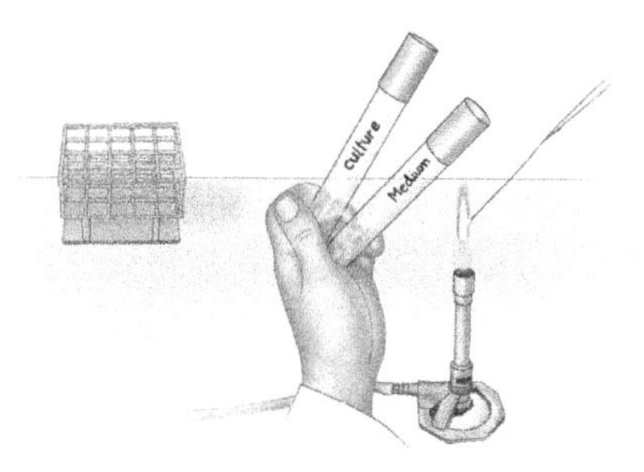

3. Remove and hold both caps with fingers on the hand holding the sterilized inoculating wire. Do not place either cap on the bench.

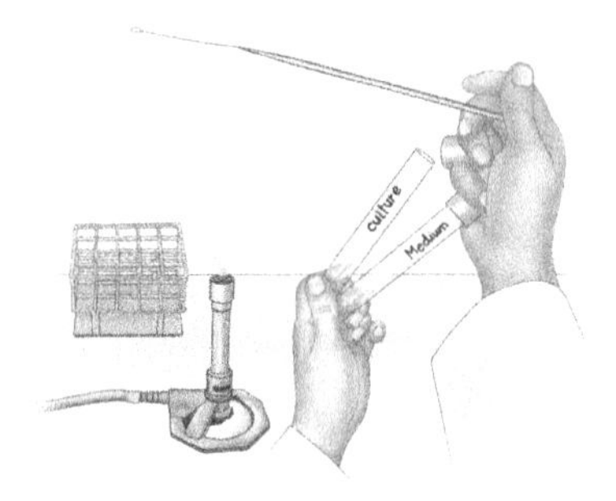

4. Flame the openings of both tubes.

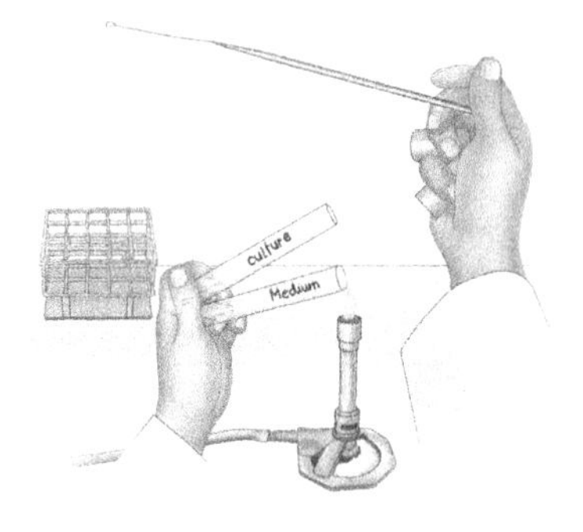

5. Insert the cooled sterile inoclating loop or needle into the tube with the culture and pick up a small inoculum of bacteria.

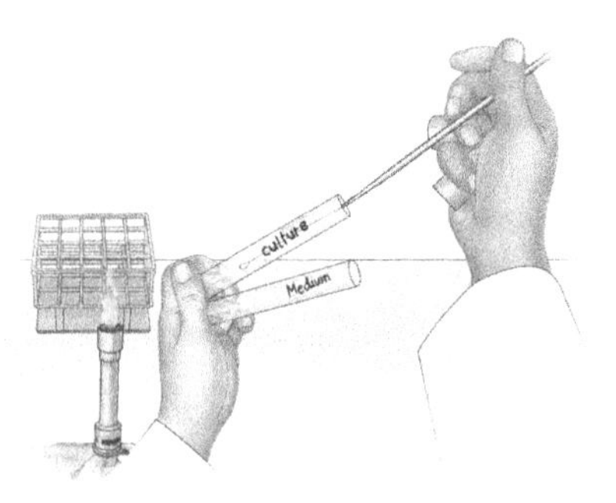

6. Inoculate the fresh tube of medium.

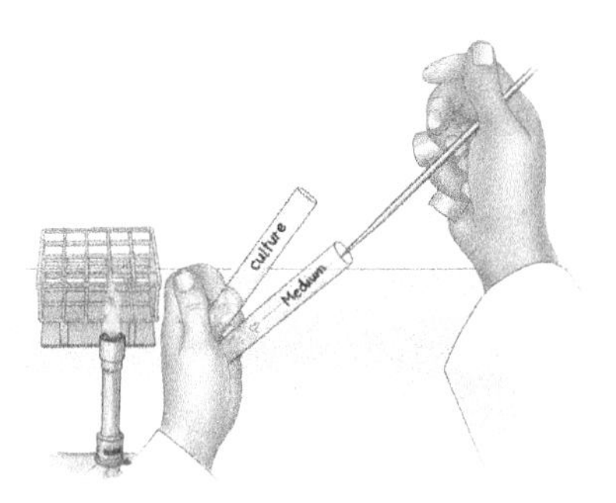

7. Flame the openings of both tubes.

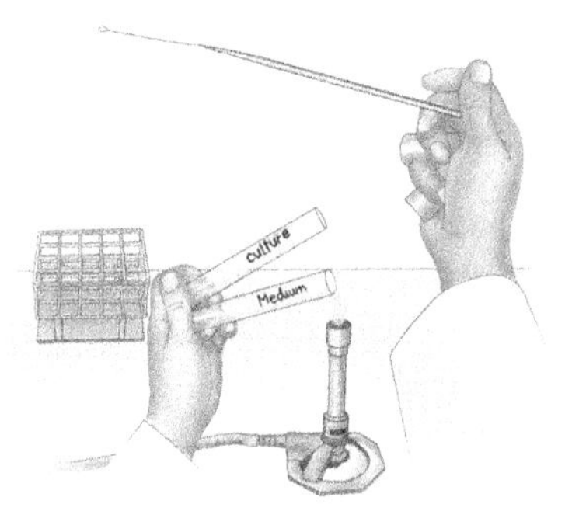

8. Replace the caps on both tubes.

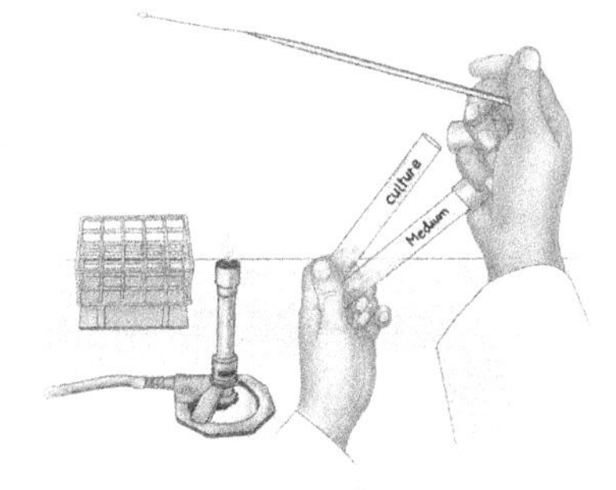

9. Flame the loop or needle until red-hot. Set it down to cool.

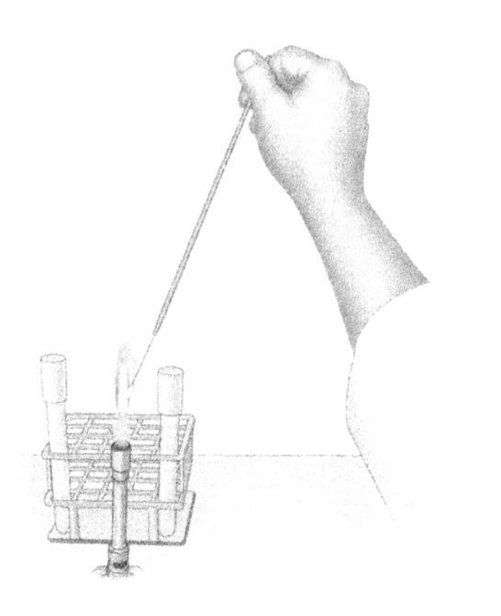

10. Place the newly inoculated tube in the incubator.

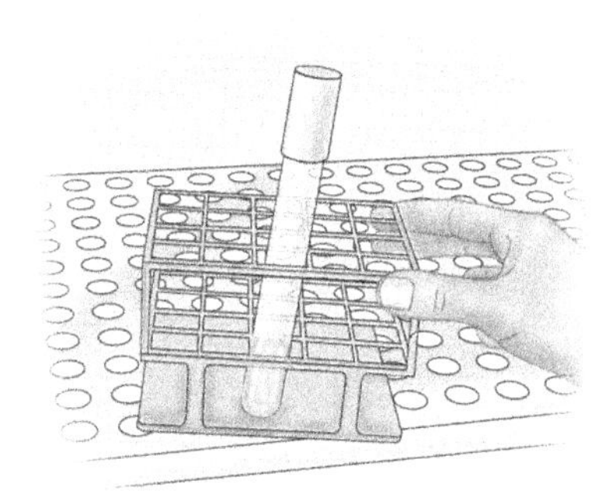

Inoculating an Agar Deep Using the Stab Technique

1. Use a flame-sterilized inoculating needle to remove an inoculum of bacteria from an agar slant.

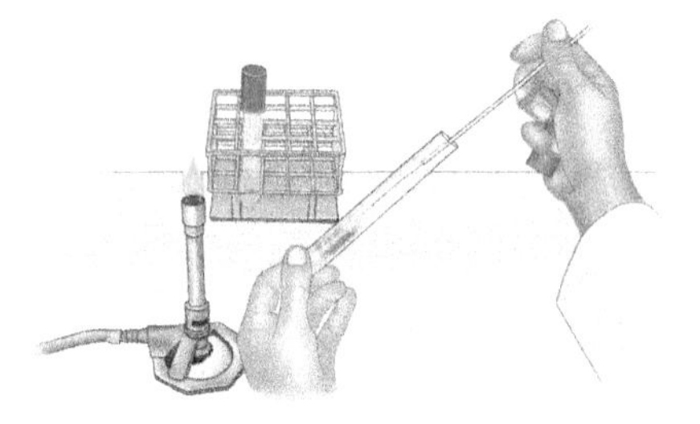

2. Remove the cap from an agar deep tube by grasping it with the little finger of the hand also holding the inoculating needle. Flame the tube.

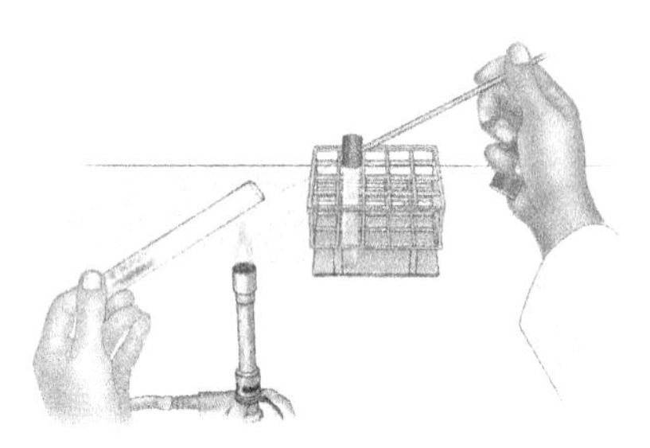

3. With the needle, "stab" the agar deep medium in the center to about two-thirds depth.

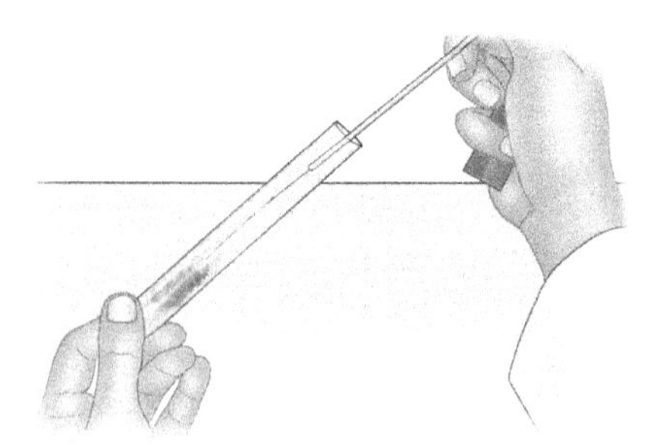

4. Withdraw the needle, flame the tube, and replace the cap.

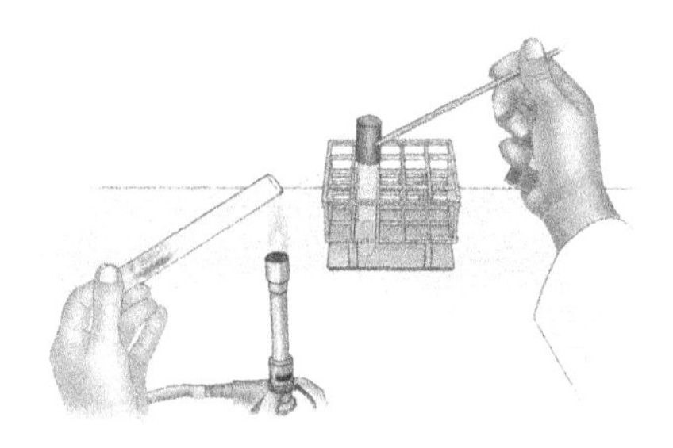

5. Flame the needle until red-hot.

6. Place the newly inoculated tube in the incubator.

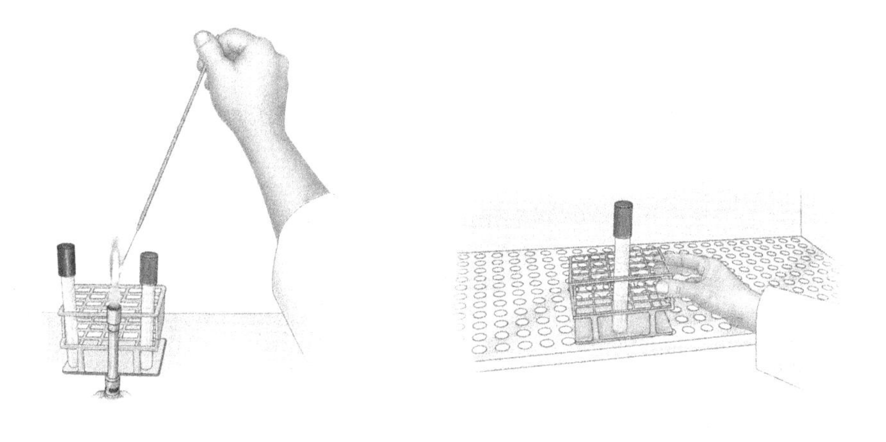

- Position the gas burner so that you do not burn yourself or your clothing.
- Tie back long hair to prevent it catching on fire.
- Turn off the gas when the flame is not needed.
- Keep culture tubes upright in a test tube rack to prevent spillage. Do not carry them in your lab coat pocket. Do not rest them against the gas jet outlet.
- Let a flamed inoculating loop or needle cool before you remove an inoculum from a culture. Spattering may create an aerosol that contains potentially harmful bacteria.
- Take care that you do not accidentally touch a hot inoculating loop or needle.

Isolation of Bacteria

Preparing a Streak Plate

Purpose:

- To isolate bacterial colonies growing on an agar plate surface in order to prepare a pure culture from a mixture of bacteria.

A **streak plate** is the technique most commonly used in the microbiology lab for isolating bacteria as a pure culture. Bacteria typically live in communities with different species. A streak plate is the first step in identifying a particular species within this mixture of species. In the procedure, the bacteria are diluted with an inoculating loop as the cells are spread over the surface of an agar plate.

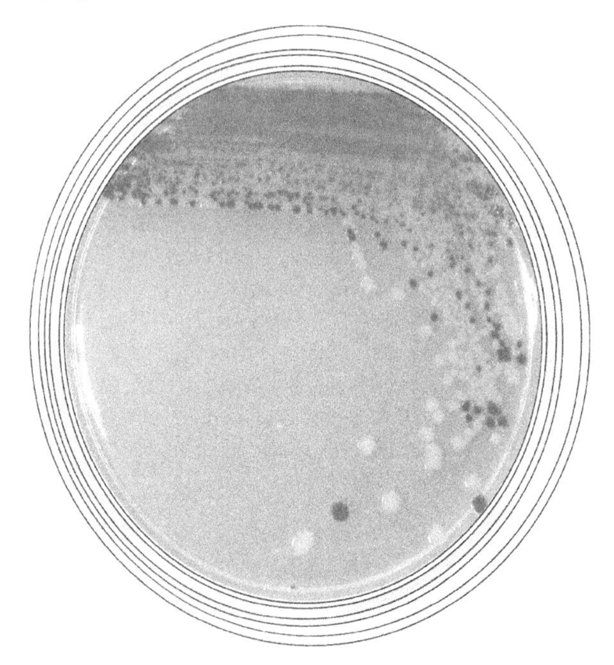

A streak plate should have well-isolated colonies so an individual colony can be removed and subcultured for characterization of the bacterium.

Materials:

- Petri plate containing growth medium agar
- mixture of bacteria in broth
- inoculating loop

Procedure:

1. Label the bottom of the plate with the source of mixed bacteria, your name or initials, and the date.

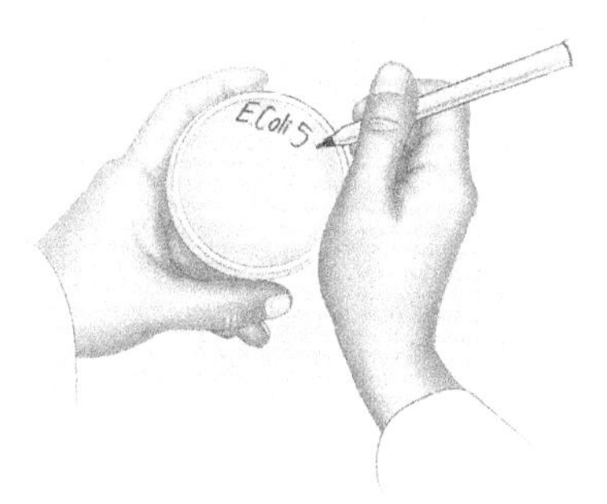

2. If the bacterial mixture is in a broth culture, suspend the cells by gently shaking the tube. Do not hold the tube by the cap: it might come off.

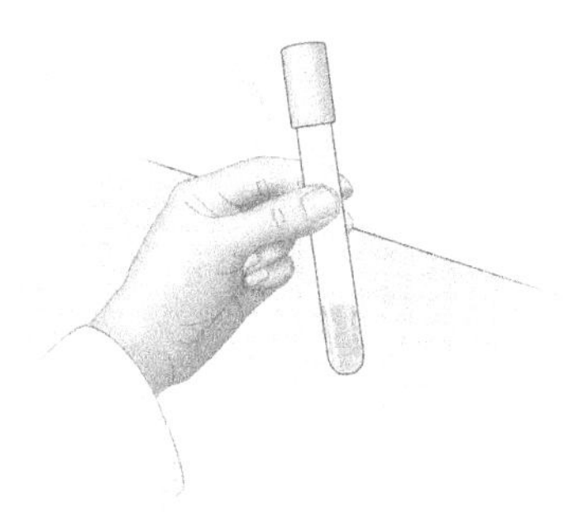

3. Hold the inoculating loop handle as you would a pencil. Sterilize the inoculating loop by flaming the wire until it is red-hot along its length. Allow the loop to cool but do not let it touch anything.

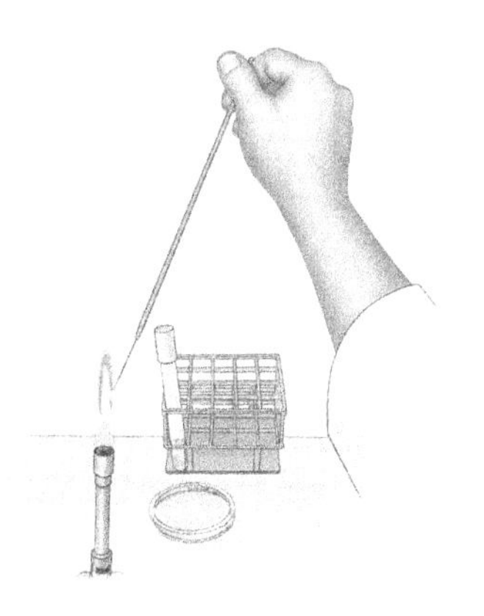

4. Hold the culture tube in your other hand. Curl the little finger of the hand holding the loop around the cap and remove it by twisting the tube with your other hand. Do not set the cap down.

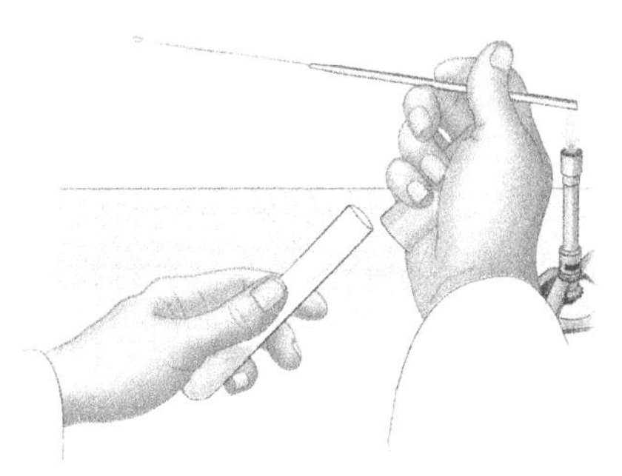

5. Quickly pass the tube mouth through the flame three times. Hold the tube at an angle to minimize the chance of dust particles falling into the open tube.

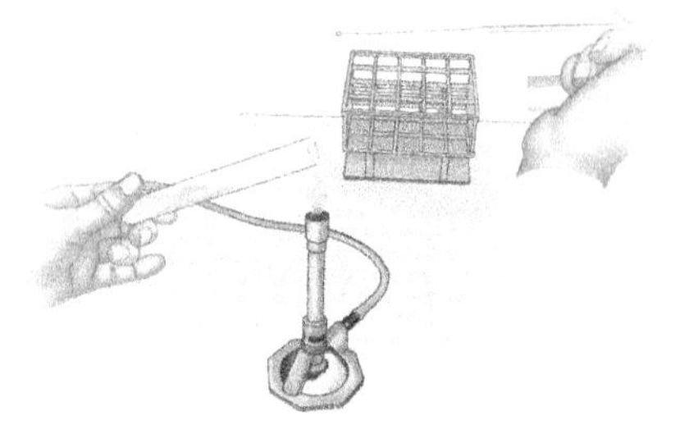

6. Insert the cooled loop into the broth culture and withdraw a bead of culture held within the loop.

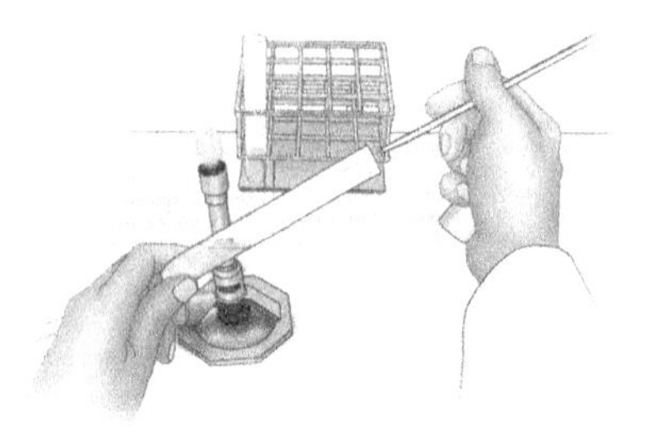

7. Pass the mouth of the tubes through the flame three times. Replace the cap on the culture tube and set it in the test tube rack. Take care that you do not dislodge the bead of culture in the loop.

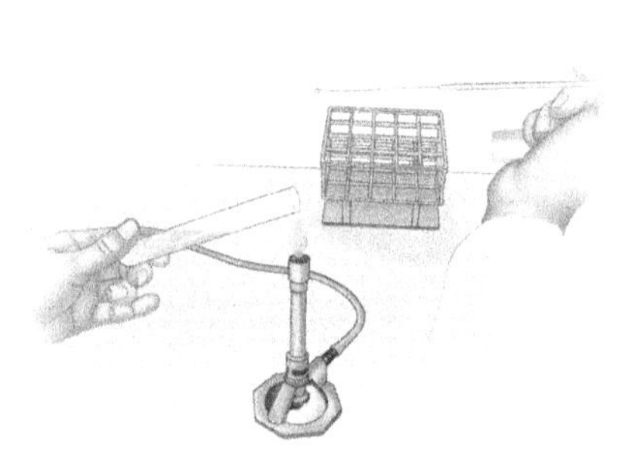

8. Lift one side of the Petri plate lid just enough so you can insert the loop and place the bead of culture on the far surface of the agar. You can rest the plate on the bench top.

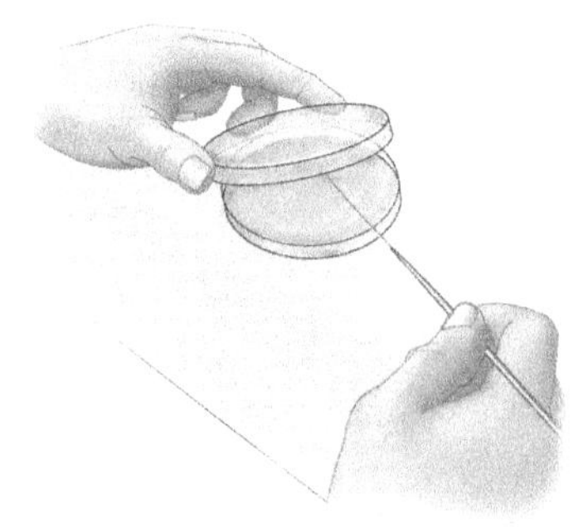

8a. Alternatively, you can hold the plate in your hand as you streak the plate.

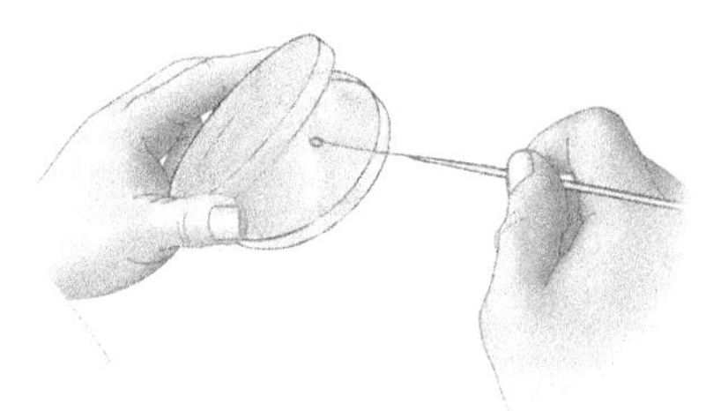

9. Streak, or sprekad out, the bacteria in the bead of culture by moving the tip of the loop in a back-and-forth motion. Do this in the first quadrant (1), as illustrated below. Then sterilize the loop by flaming it. After the loop cools, start the second quadrant by moving the loop tip through the last few streaks in the first quadrant. Repeat for the remaining two quadrants.

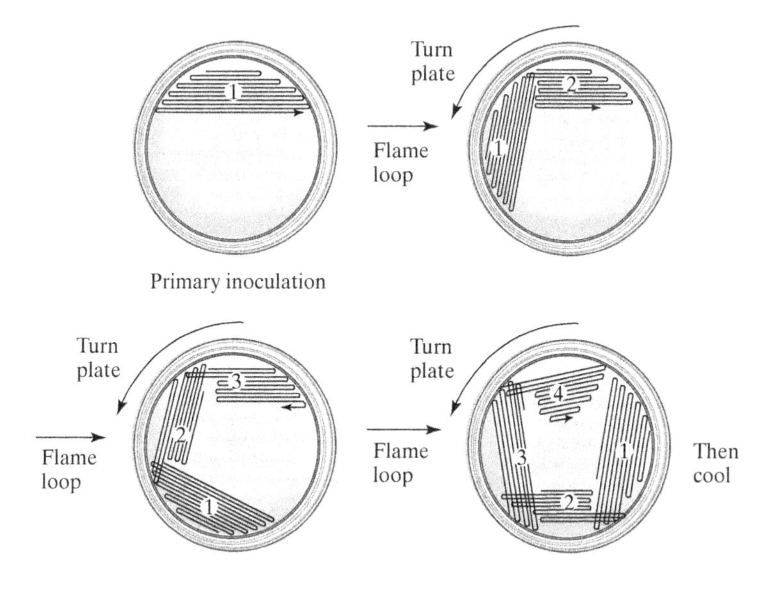

10. When you have finished preparing the streak plate, flame the loop before you set it down.

11. Place the plate upside-down (agar side up) in the incubator at 37° C (or at a temperature appropriate for the bacteria) for at least 24 hours.

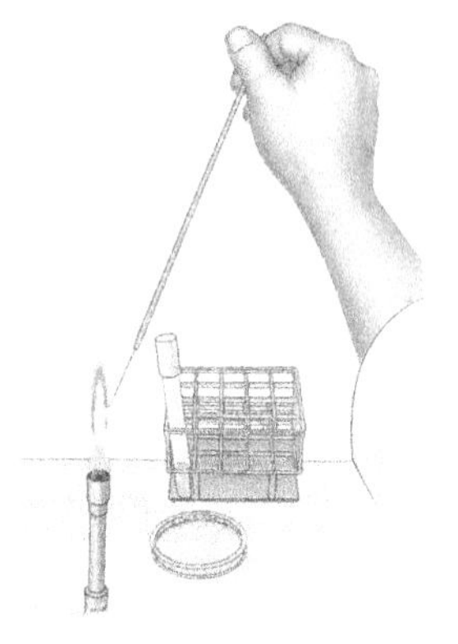

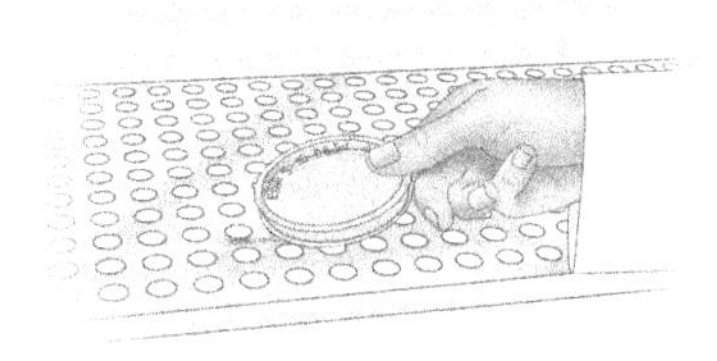

12. After incubation, look for isolated colonies on the agar surface. Use a flame-sterilized inoculating loop to remove an isolated colony.

13. Use aseptic technique to transfer the inoculum on the loop to an agar slant tube (pp. 35–37) or a new agar plate. Incubate the tube or plate.

14. When you have finished with the streak plate, discard it in the designated place for sterilization.

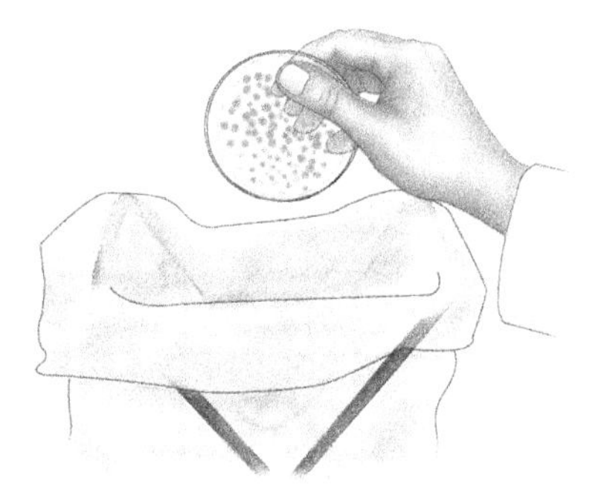

Troubleshooting

- When you streak an agar plate, the inoculating loop must be bent at a slight angle. This will minimize gouging the agar as you streak with the loop tip over the agar surface.
- Use a gentle gliding stroke. Too much pressure increases the likelihood of gouging.
- Place the plate on the lab bench or hold the plate so you can observe the lines that form on the agar surface as you move the loop back and forth.
- If you do not flame the loop after you streak a quadrant, the bacterial cells will not be as diluted as when you flame between each quadrant.
- Allow the flamed inoculating loop to cool before you streak a quadrant; otherwise the bacteria you are trying to spread will die from the heat. You can cool the loop at the edge of the agar surface.
- If you transfer too many bacteria from one quadrant to the next while streaking, the cells may not be adequately separated to grow into distinct colonies.

ISOLATION OF BACTERIA

Preparing a Pour Plate

Purpose:

- To isolate bacterial colonies in agar or on its surface in order to prepare a pure culture from a mixture of bacteria.

A **pour plate** is useful for isolating bacteria by diluting a mixture of different species. In the pour-plate technique a loopful of a suspension of mixed bacteria is diluted in a short series of melted agar tubes. The contents of each tube are poured into empty Petri plates. One of the dilutions will yield separated colonies on the agar surface or within the agar. These colonies can then be transferred to an agar slant. The growth on the slant will be a pure culture.

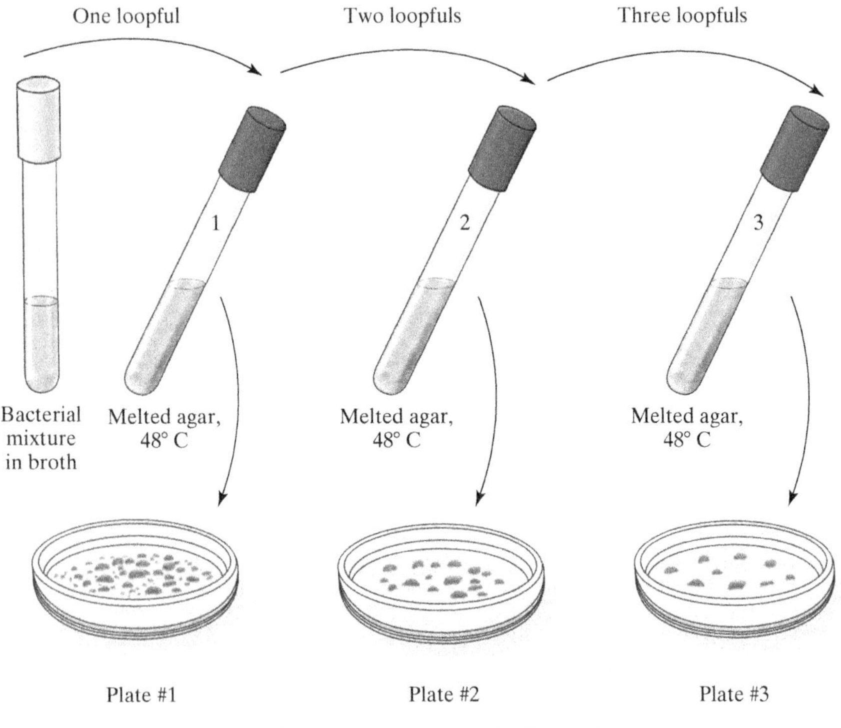

The pour plate technique for isolating bacteria

Materials:

- tubes containing melted trypticase soy or nutrient agar, 18 mL, kept in a water bath at 48–50° C

  or

  - tubes containing trypticase soy or nutrient agar, 18 mL
  - beaker, 250 mL or 400 mL, half-filled with water
  - thermometer
  - hot plate
- sterile empty Petri plates
- inoculating loop
- mixture of bacteria in a broth tube

Procedure:

1. Label the bottoms of three empty Petri plates with your name or initials and the date. Label the plates 1, 2, and 3.

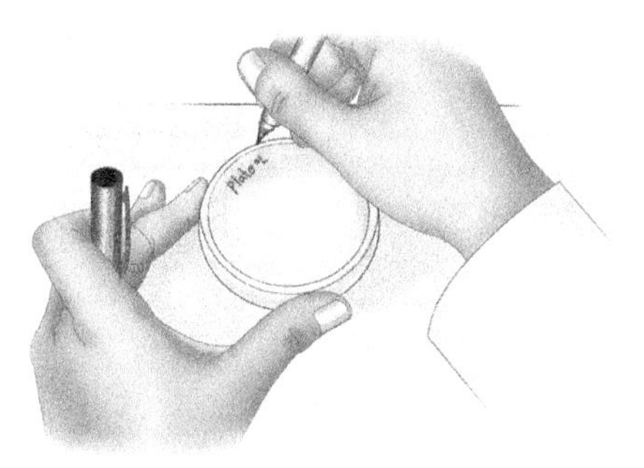

2. Label three agar deep tubes 1, 2, and 3. If the agar is not already melted, heat the tubes in a beaker with water on a hot plate until the agar is liquefied. Cool the tubes to about 48–50° C.

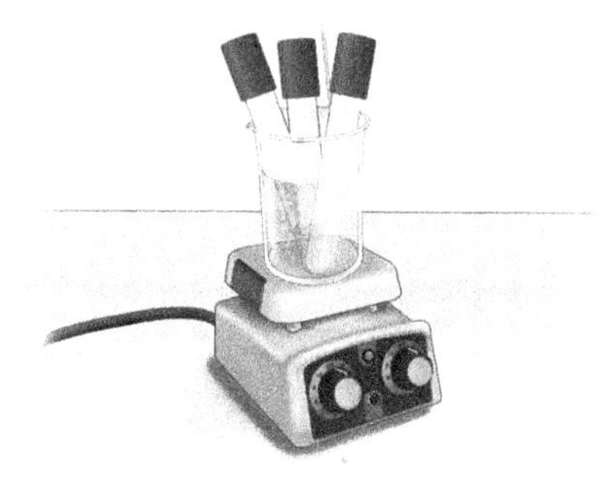

3. Suspend bacterial cells in the broth mixture by gently shaking the tube. Do not hold the tube by the cap: it might come off.

4. Remove tube 1 from the hot water bath. Hold it in front of the culture tube.

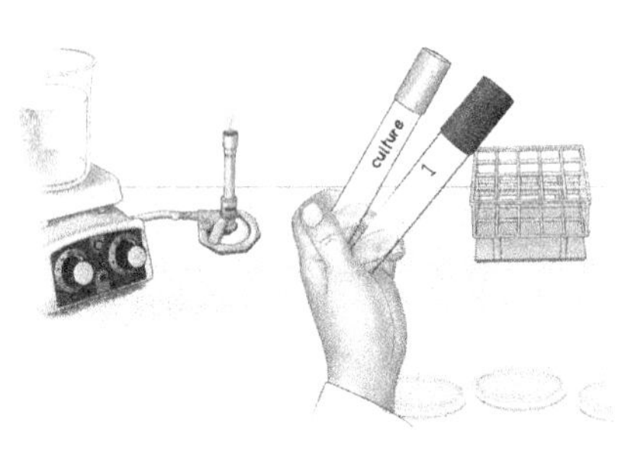

5. Hold the inoculating loop handle as you would a pencil. Sterilize the inoculating loop by flaming the wire until it is red-hot along its length. Allow the loop to cool but do not let it touch anything.

6. Remove the caps on the tubes with the fingers of the hand holding the loop. Quickly pass the mouths of the tubes through the flame three times.

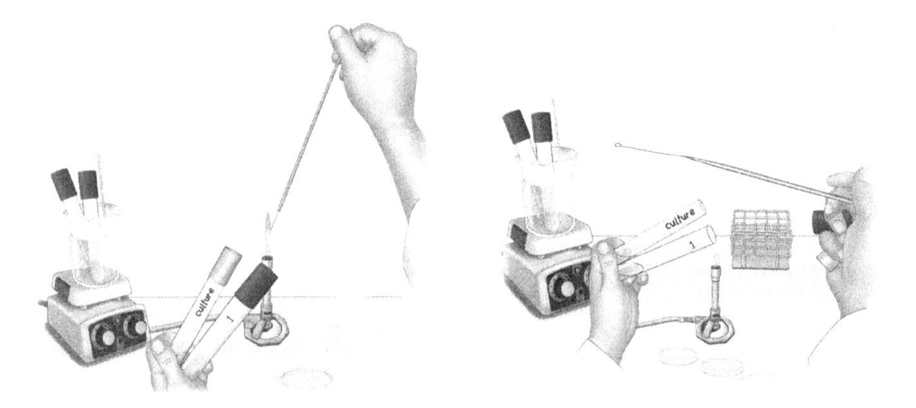

7. Inoculate tube 1 with a loopful of broth culture. Twirl the handle to ensure complete transfer of the bacteria.

8. Pass the mouths of the tubes through the flame three times. Replace the caps. Place the culture tube in the test tube rack. Sterilize the loop in the flame and set it down.

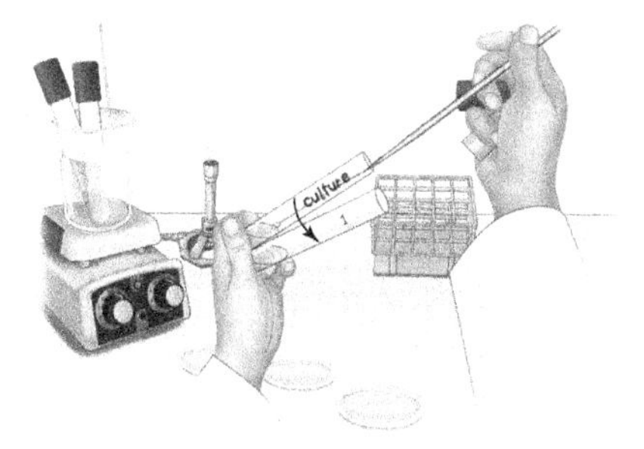

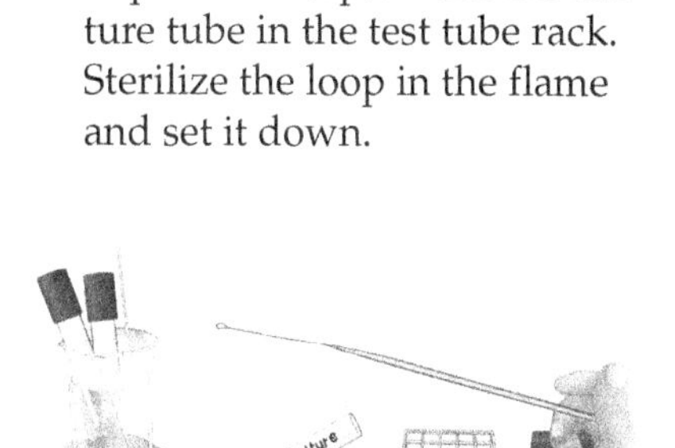

9. Mix the contents in tube 1 by vigorously rolling the tube between your palms about 25 times.

10. Remove tube 2 from the hot water bath. Hold it in front of tube 1.

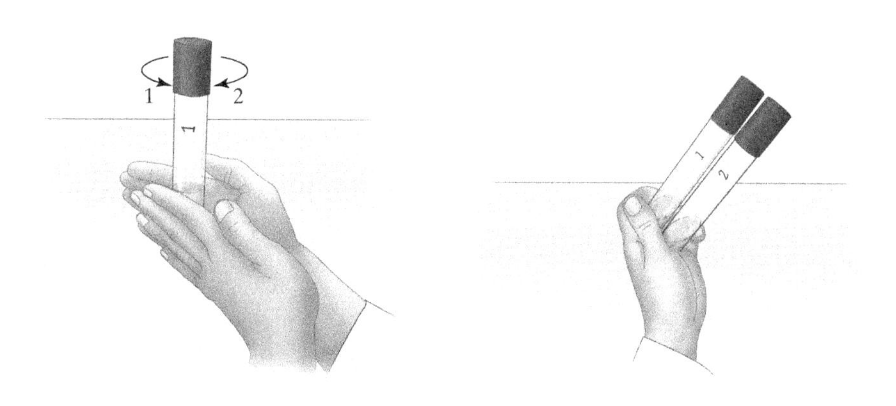

11. Aseptically transfer two loopfuls of bacteria/melted agar from tube 1 into tube 2. Twirl the handle to ensure complete transfer.

12. Pass the mouths of the two tubes through the flame three times. Replace the caps. Return tube 1 to the hot water bath. Sterilize the loop in the flame and set it down.

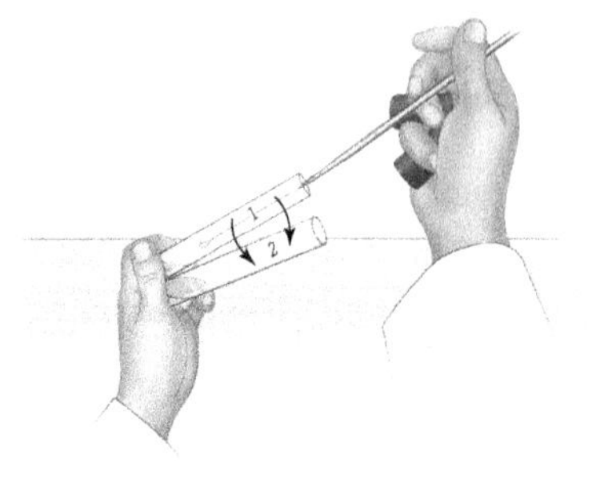

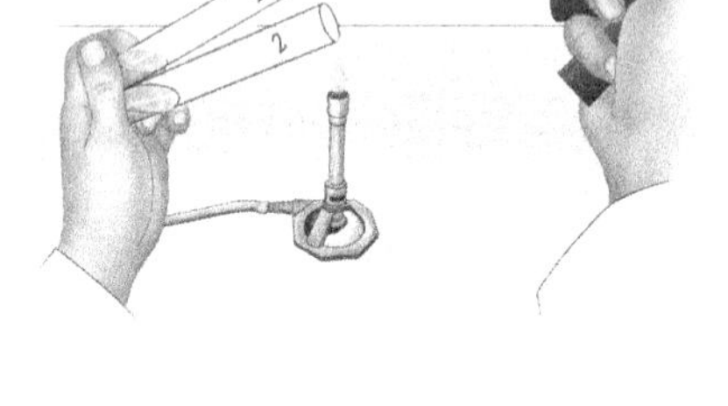

13. Mix the contents in tube 2 by vigorously rolling the tube between your palms about 25 times.

14. Remove tube 3 from the hot water bath. Hold it in front of tube 2.

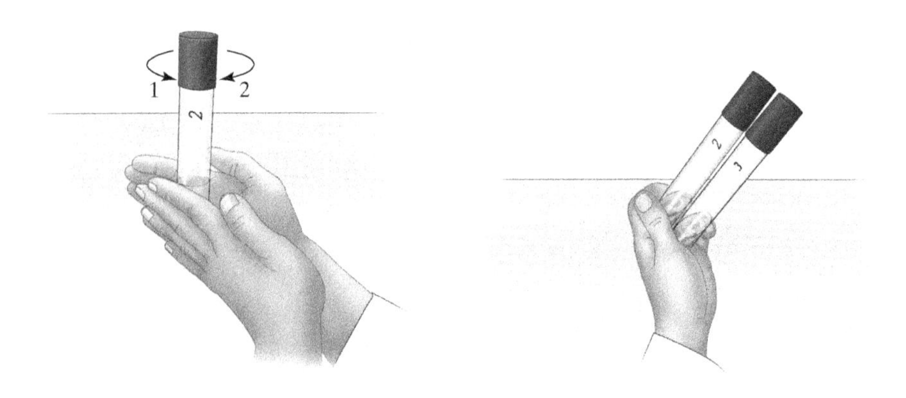

15. Aseptically transfer three loopfuls of bacteria/melted agar from tube 2 into tube 3.

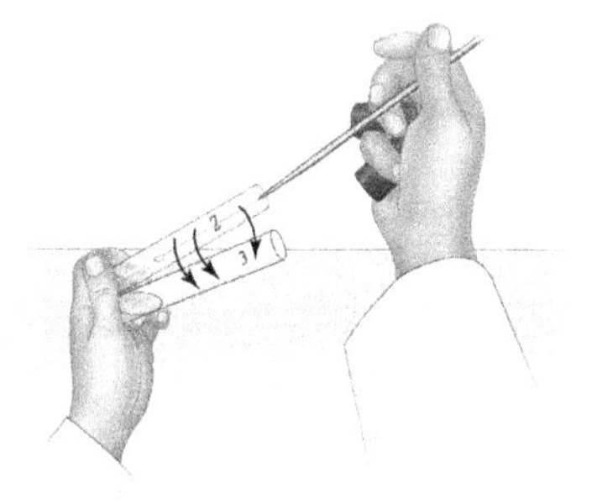

16. Pass the mouths of the tubes through the flame three times. Replace the caps. Return tube 2 to the hot water bath. Sterilize the loop in the flame and set it down.

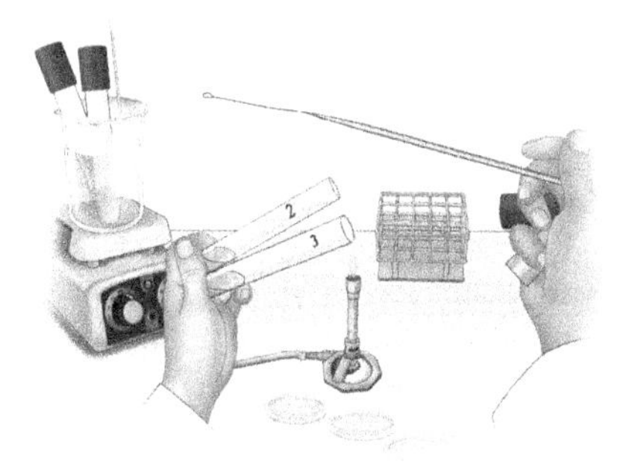

17. Mix the contents in tube 3 by vigorously rolling the tube between your palms about 25 times.

18. Use aseptic technique to pour the contents of each tube into its respective plate.

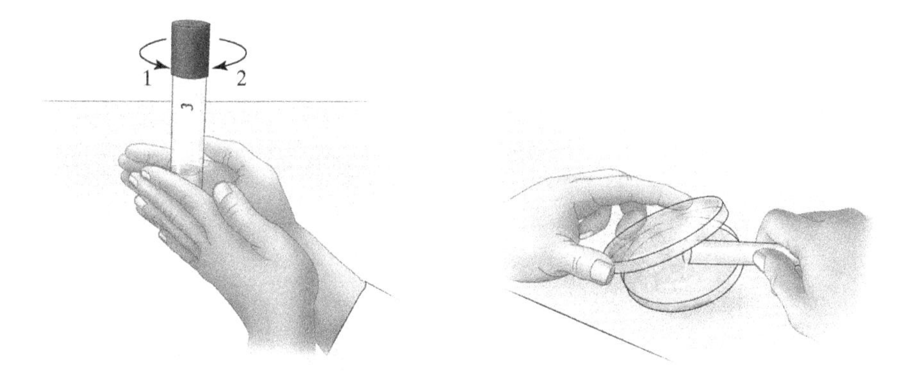

19. Spread the melted agar over the bottom of each plate by gentle swirling. Avoid splashing over the edge or onto the lid.

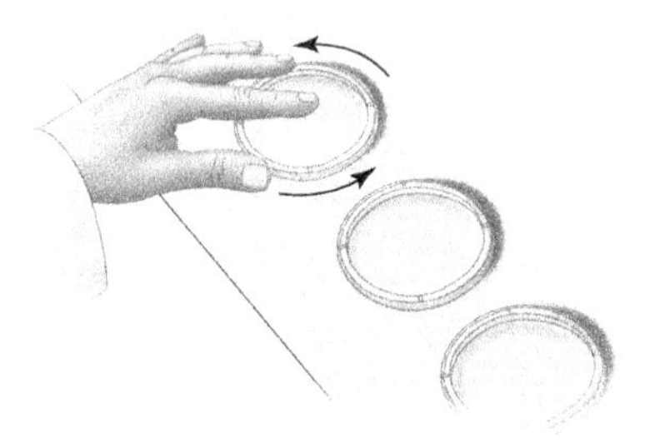

20. After the agar has solidified, place the plates upside-down (agar side up) in the incubator at 37° C (or at a temperature appropriate for the bacteria) for at least 24 hours.

21. After incubation, transfer isolated colonies to agar slants or plates. Use a sterile inoculating needle to pick up a colony embedded in agar and/or use a sterile inoculating loop to pick up a colony on the agar surface.

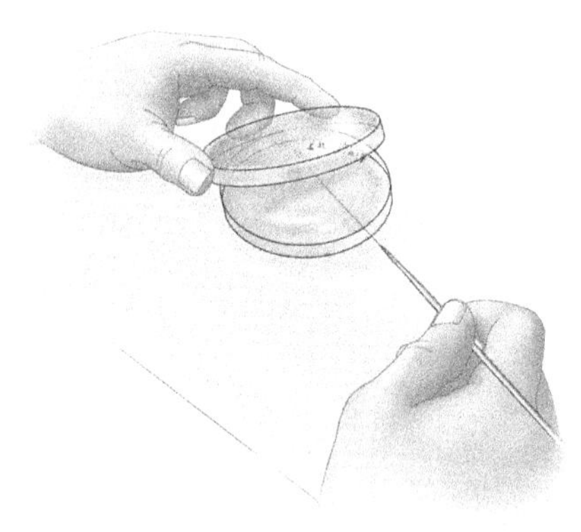

22. Incubate the agar slant or plate for at least 24 hours. Test the growth for a pure culture.

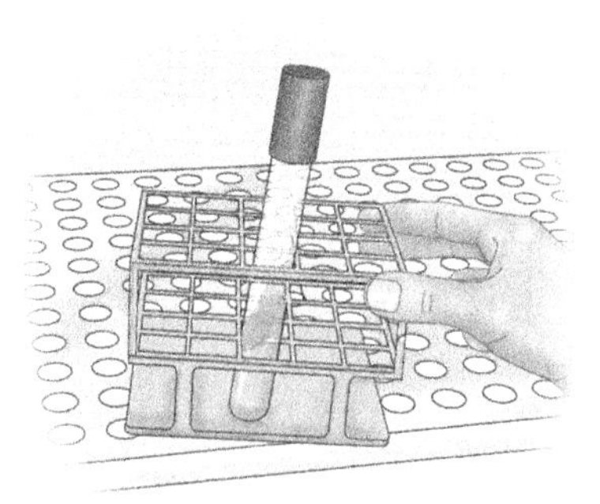

23. When you have finished with the pour plates, discard them in the designated place for sterilization.

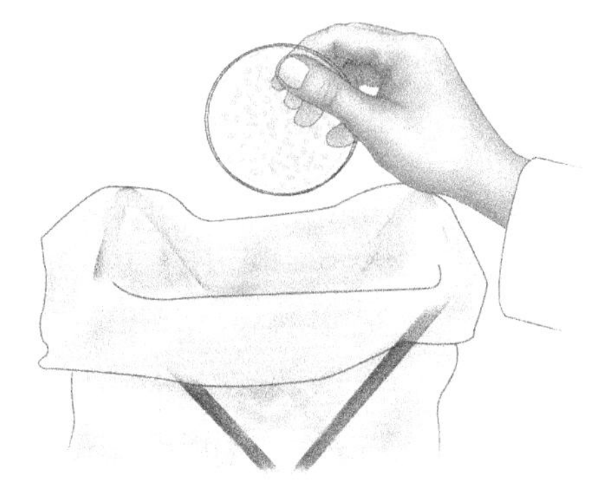

Troubleshooting

- Only bacteria that can withstand brief exposure to the temperature of the melted agar can be isolated by the pour plate technique.
- Allow the agar tubes that were melted in boiling water to cool, but make sure the agar remains liquid in order to avoid premature gelling and unwanted bacterial death. It is best to keep melted agar tubes in a hot water bath until you need to use the liquefied agar.
- Work quickly so that the melted agar in the tube does not solidify before you pour the mixture into a sterile Petri plate.
- Avoid vigorous swirling of the melted agar to avoid splashing over the edge or onto the lid of the plate.
- Pour melted agar carefully but quickly into the plate to prevent bubbles.
- Agar will become translucent and lighter in color when it has solidified.

Isolation of Bacteria

Preparing a Spread Plate

Purpose:

- To isolate bacterial colonies on an agar plate surface in order to prepare a pure culture from a mixture of bacteria.

In a **spread plate**, a small volume of broth containing a mixture of bacteria is spread over the surface of an agar plate. This technique is most successful if the concentration of bacteria in the broth mixture is low.

Materials:

- 150 mL beaker with 60 mL of 95% ethanol
- glass spreader *or* sterile disposable plastic spreader
- Petri plate with agar medium
- sterile 1.0 mL pipette
- pipettor
- mixed bacterial culture in broth

Procedure:

1. Label the bottom of the plate with the source of mixed bacteria, your name or initials, and the date.

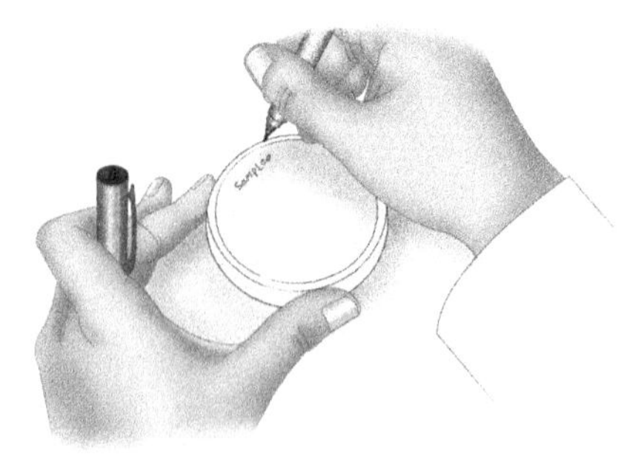

2. Set up the materials as shown below. Position the beaker containing alcohol away from the open flame.

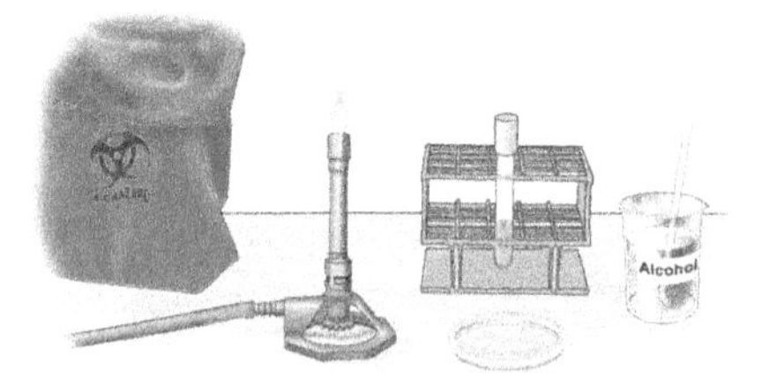

3. Use aseptic technique to remove 0.2 mL of broth culture.

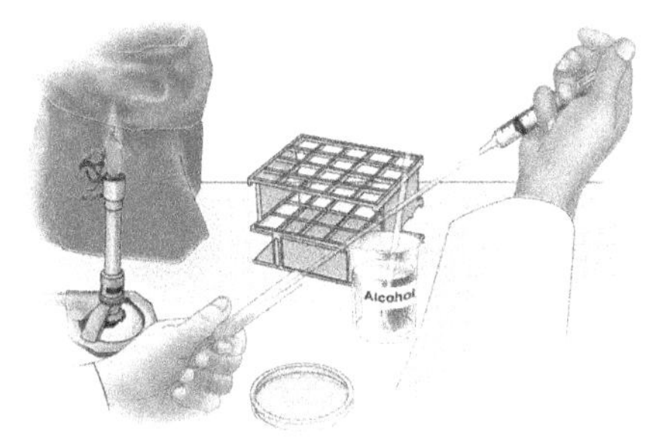

4. Lift the lid of the plate just enough to allow you to place the pipette tip in the center. Dispense 0.1 mL. Return the remaining broth culture to its tube. Place the pipette in the designated container for sterilization.

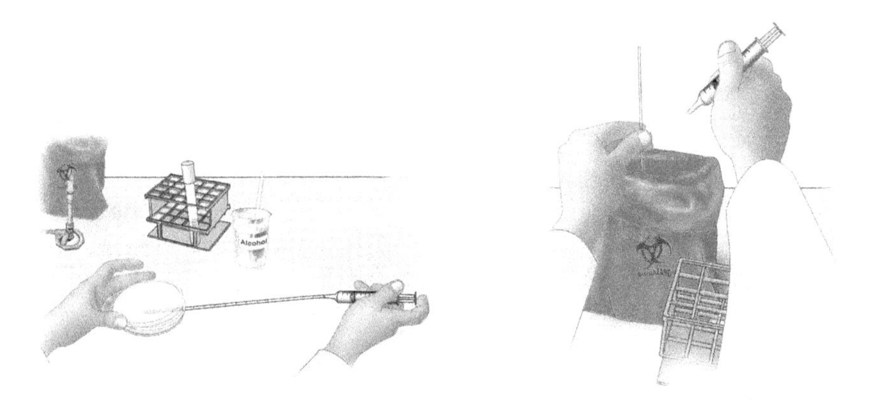

5. If you are using a glass spreader, complete Steps 5–9 within 15–20 seconds. If you are using a sterile plastic disposable spreader, begin at Step 8.

6. Dip the glass spreader into the alcohol. Ignite the alcohol on the spreader. Do not keep the spreader in the flame.

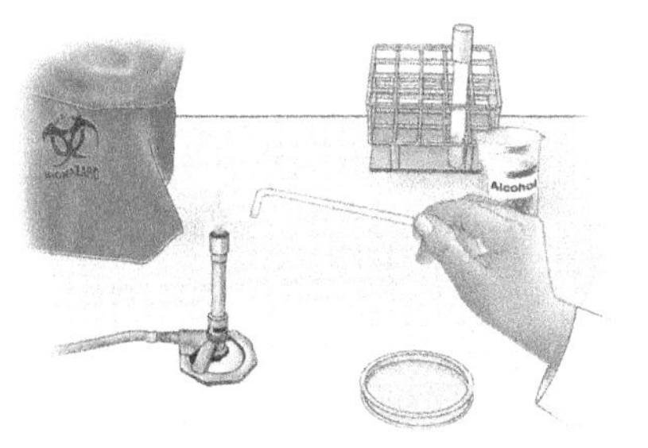

7. As soon as the alcohol burns off, lift the lid and touch the spreader at the edge of the agar to cool the glass.

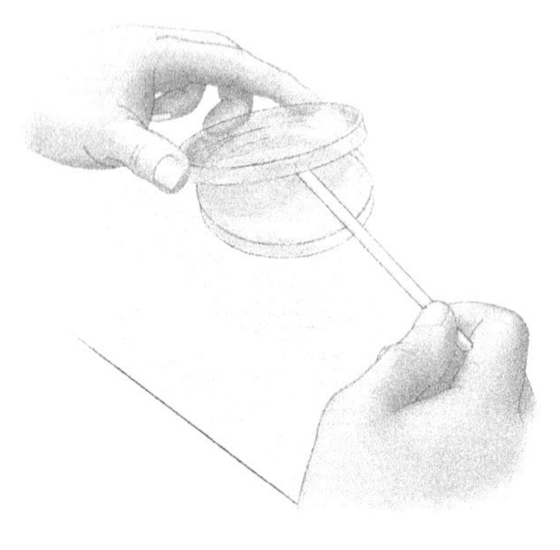

8. Touch the cooled spreader to the sample in the center of the agar. Use the spreader to push the sample over the entire agar surface.

8a. As an alternative, you can spin the plate on a special turntable designed for this.

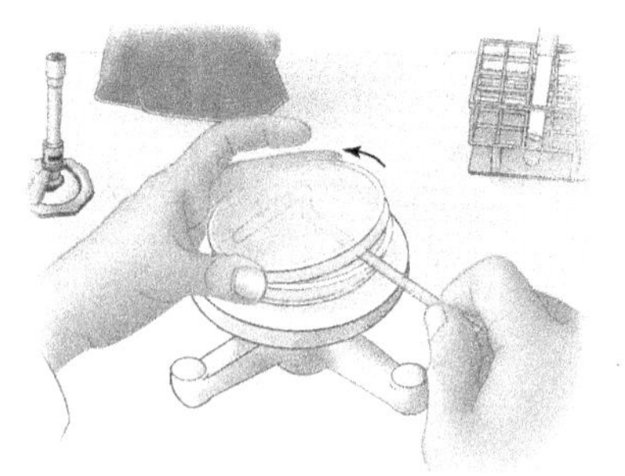

9. After you replace the lid, dip the glass spreader in the alcohol. Ignite the alcohol and allow it to burn off.

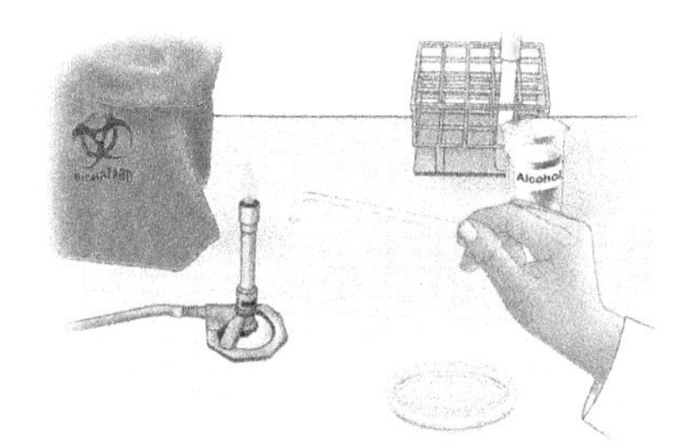

9a. If you are using a plastic spreader, dispose of it in a biohazard bag.

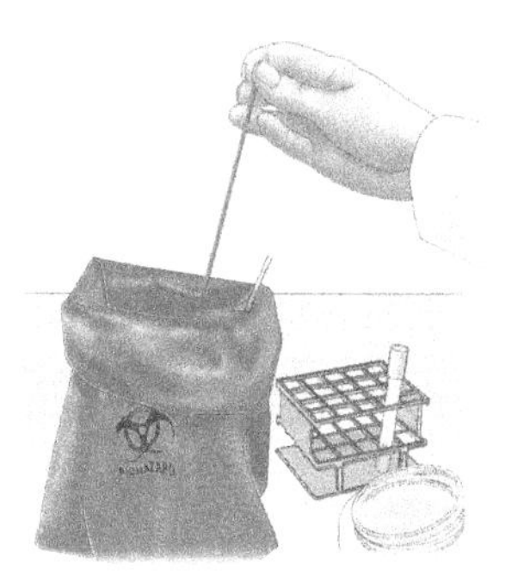

10. Place the plate upside-down in the incubator for 24–48 hours.

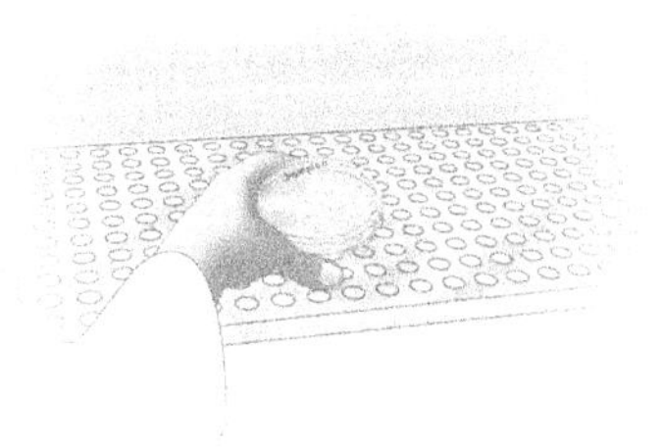

11. After incubation, culture an isolated colony on an agar slant (pp. 35–37).

12. When you have finished with the spread plate, discard it in the designated place for sterilization.

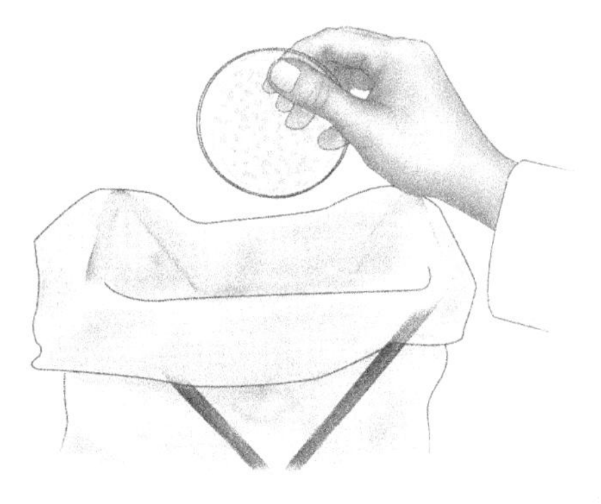

- *Position the beaker containing alcohol away from the open flame.*
- *Take care so drops of flaming alcohol do not fall into the beaker or onto flammable materials on the lab bench surface. If the alcohol in the beaker is ignited, cover the beaker immediately to block air and so douse the flames. Know the loca tion of the nearest fire extinguisher.*
- *Use a mechanical pipettor. Never use your mouth to draw liquids into a pipette.*

Troubleshooting

- A sterile pipette or disposable spreader should be removed from its plastic wrapping just before it is used.

Characteristic Features of Bacterial Growth in Culture

Bacterial Colonies Growing on Agar Plates

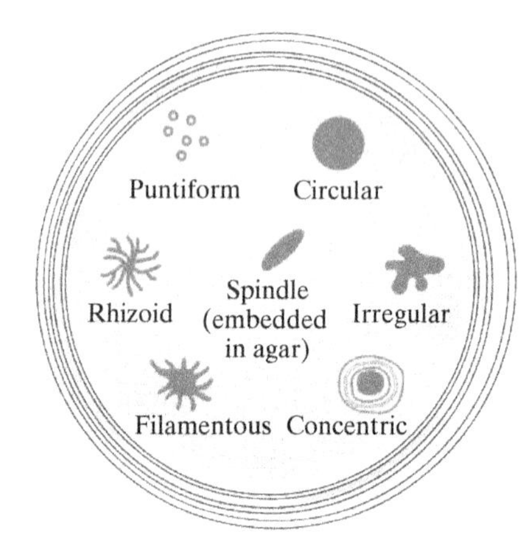

Whole colony (form)

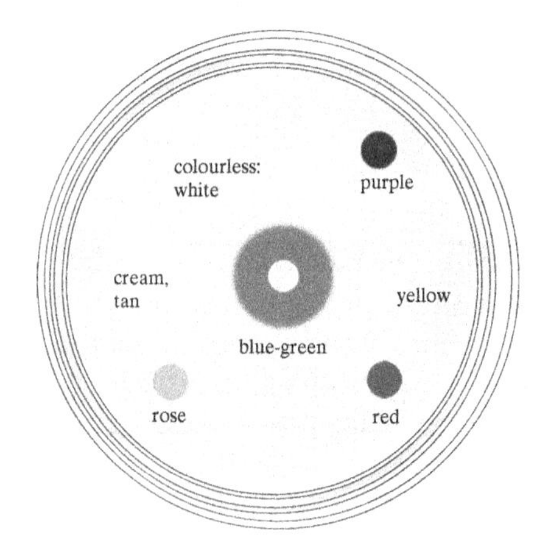

Colony pigmentation

Appearance of colony surface

- opaque
- transparent
- translucent
- rough or dull
- smooth or glistening

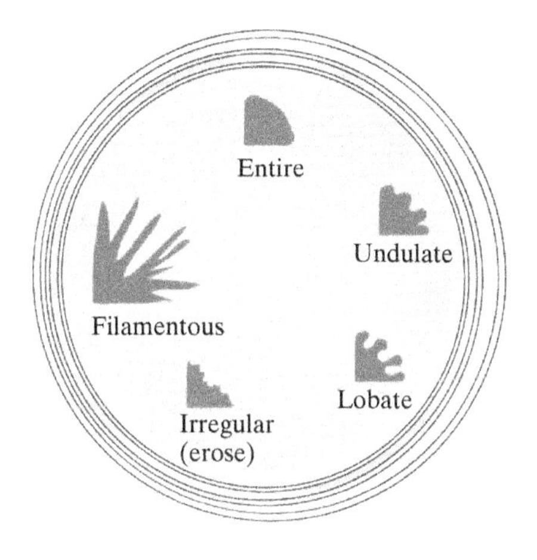

Colony margin (edge)

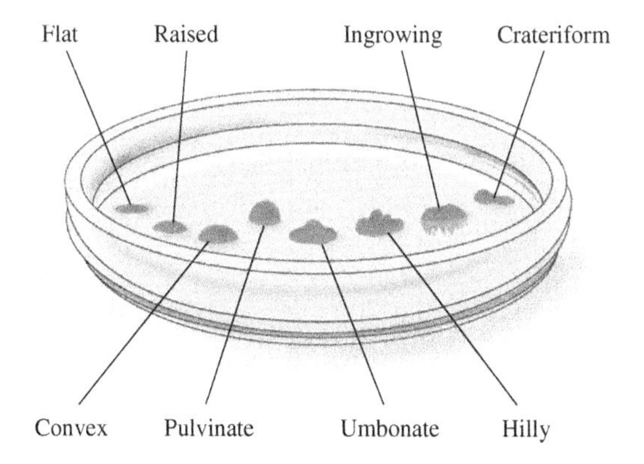

Elevation of colony

Bacteria Growing on an Agar Slant

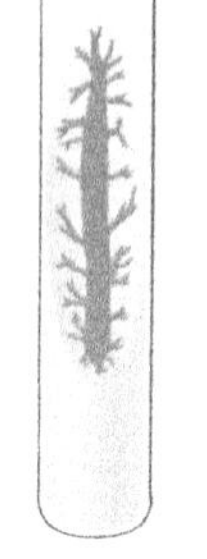

Arborescent (branched)

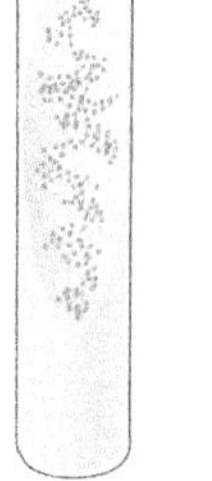

Beaded

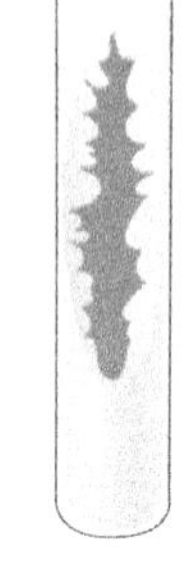

Echinulate (pointed)

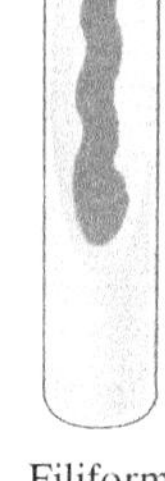

Filiform (even)

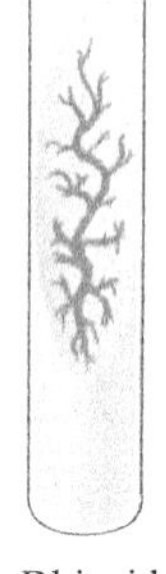

Rhizoid (rootlike)

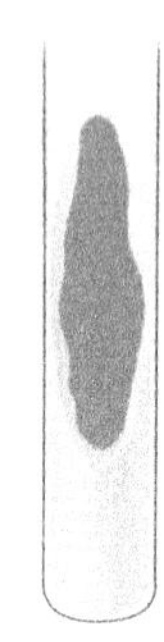

Spreading

Bacteria Growing in a Broth Medium

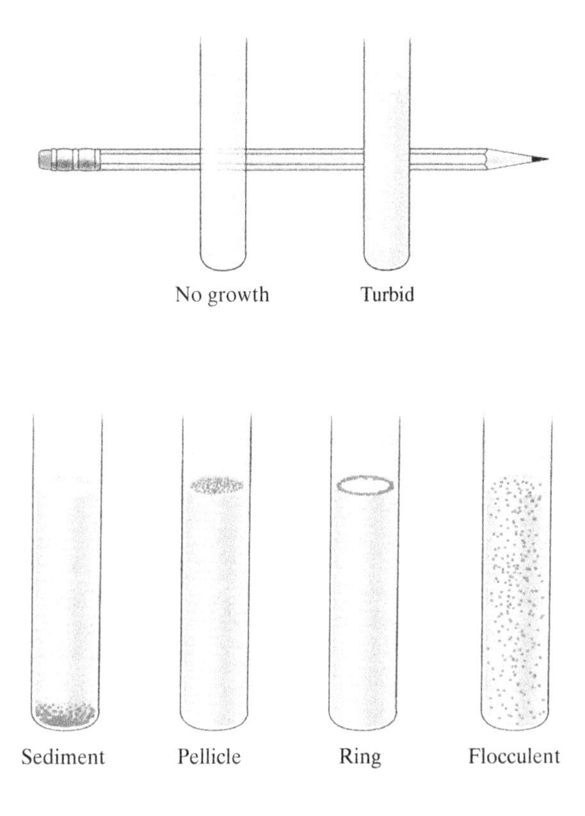

Maintenance and Storage of Stock Cultures

Purpose:

- To maintain stock bacterial cultures for day-to-day use.
- To preserve stock cultures for long-term storage.

Short-term Storage of Stock Cultures

Two types of stock cultures are set up in the laboratory and stored in a refrigerator (4° C) for short periods of time. For daily work, a **working stock culture** is maintained on a trypticase soy or nutrient agar slant that is stored in the refrigerator. After several days of use or accidental contamination, a fresh working culture is prepared with an inoculum taken from a **reserve stock culture**, also a slant culture, stored in the refrigerator. A reserve stock culture is best kept in a tube with a screw cap to prevent the medium from drying out. A fresh reserve culture should be prepared each month.

Long-term Storage of Stock Cultures

Two procedures are available for storing stock cultures for long periods of time: **freeze-drying (lyophilization)** and **freezing**.

Many laboratories lack the heavy-duty vacuum pump and chambers needed for lyophilization; that technique will not be described here.

Because freezing can cause formation of mini-ice crystals that can damage microbial cell membranes, a cryoprotectant, such as glycerol, is included in the storage medium. While freezing cryoprotected cells at very low temperatures (e.g., –196° C in liquid N_2) will preserve the cells for a very long time, cryoprotected microbes can be stored at –20° C or at –70° C. Many bacteria stored at –20° C can be revived from cold storage for 1–2 years, after which they must be recultured and prepared for another round of freezing. At –70° C, non-fastidious bacteria maintain viability for about 5 years and fastidious bacteria for 3 years.

Materials:

- sterile 1.5 mL microcentrifuge tube (autoclaved)
  or
  sterile cryovial, 2.0 mL, outside threads
- sterile glycerol (autoclaved)
- micropipettor, 100–1000 µL
- micropipette tips, sterile, 100–1000 µL
- pure culture grown overnight in broth medium
- freezer (–20° C or –70° C [preferred])

Procedure:

1. Label the microcentrifuge tube or cryovial with the name of the bacterium and the date.

2. Use aseptic technique to transfer 150 μL of sterile glycerol to the microcentrifuge tube or cryovial. Discard the micropipette tip in a biohazard bag. Then add 850 μL of the overnight broth culture. Discard the micropipette tip in a biohazard bag.

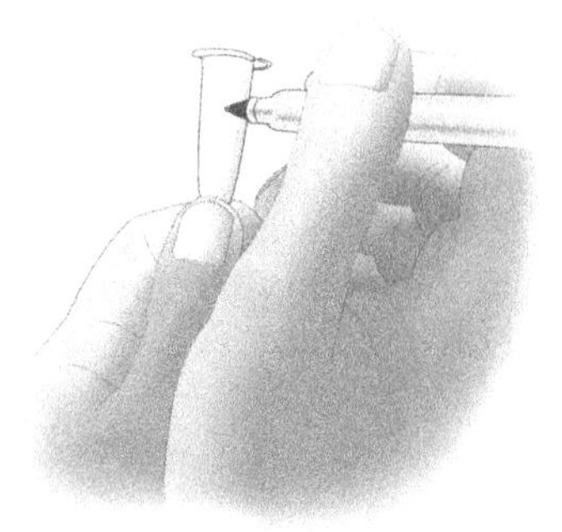

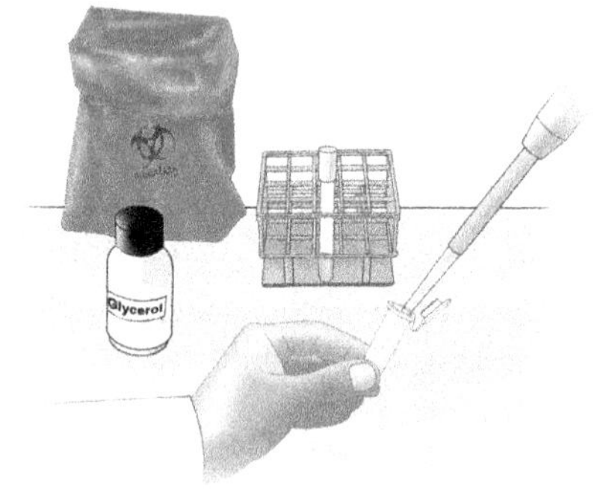

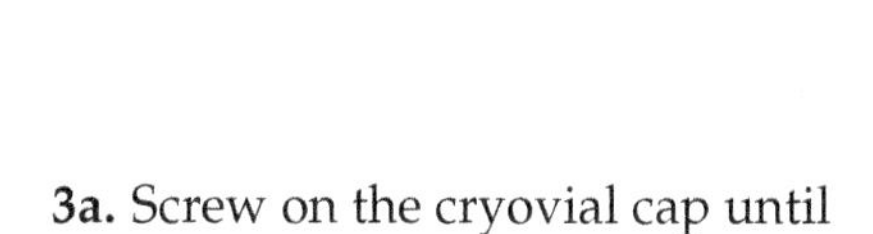

3. Firmly press down on the microcentrifuge cap for a secure fit.

Or

3a. Screw on the cryovial cap until it is finger tight.

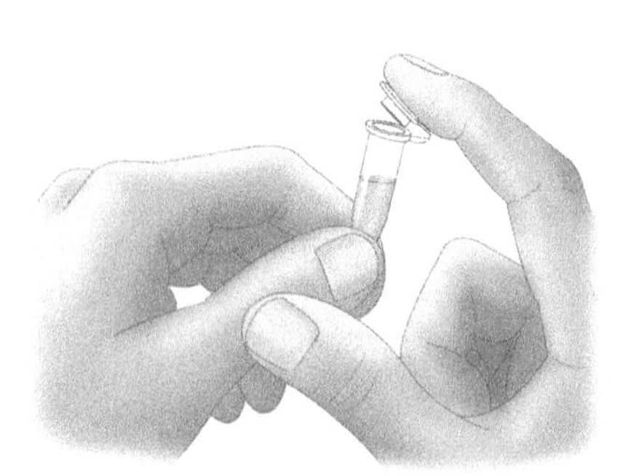

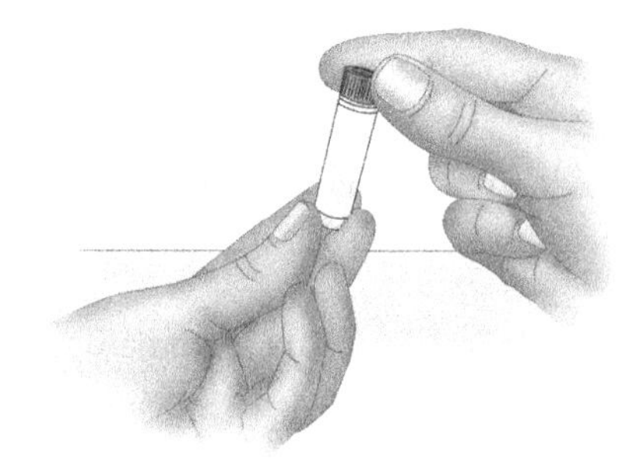

4. Mix the tube or vial contents by gentle vortexing for 10 seconds or by vigorously turning it end over end for about 20 seconds.

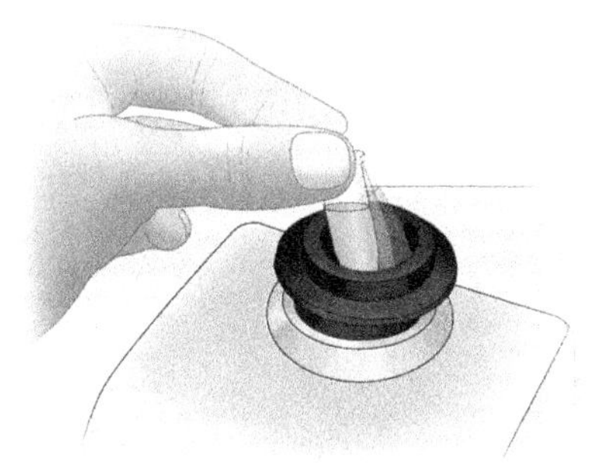

5. Place the tube or vial in a cyrostorage box. Store the box in a freezer, preferably at –70° C.

Growing a Fresh Culture from Frozen Stock

Materials:

- stock culture stored in cryovial or microcentrifuge tube in freezer
- sterile toothpicks in a 100 mL beaker covered loosely with aluminum foil (previously sterilized by autoclaving)
- Petri plate with agar medium
- biohazard bag

Procedure:

1. Bring your materials to the freezer so the stored culture has no chance to thaw. Remove the cap from the vial or lift the lid of the tube containing the glycerol stock. Use a toothpick to scrape an inoculum from the top of the frozen layer.

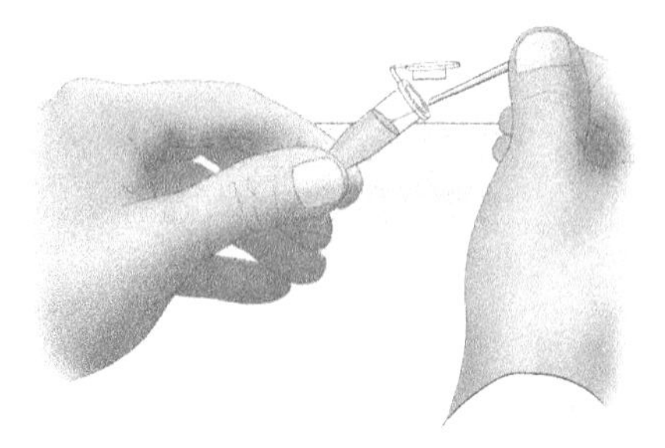

2. Transfer the material on the toothpick to the surface of the agar. Discard the toothpick in a biohazard bag. Immediately return the vial or tube to the freezer.

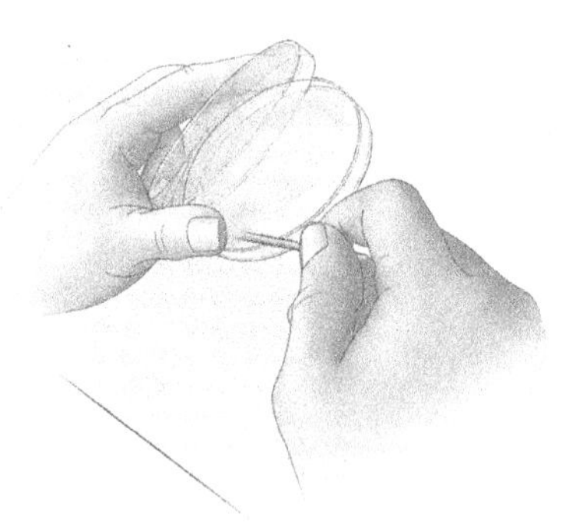

3. Prepare a streak plate of the inoculum (pp. 42–47). Incubate the plate overnight at 35° C.

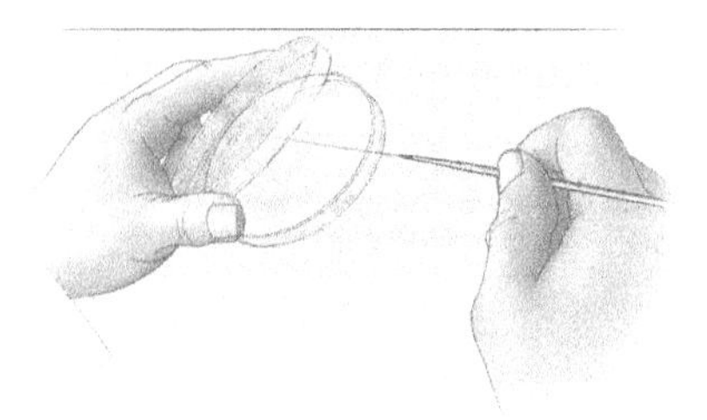

An alternative freezing and recovery procedure uses craft beads in a vial containing glycerol and growth broth (Pro-Lab Diagnostics, Austin, TX [*Microbank*] and Key Scientific Products, Round Rock, TX [*CryoCare*]).

Materials:

- streak plate with pure culture, grown overnight at 35° C
- *Microbank* or *CryoCare* vial
- sterile Pasteur pipette and rubber bulb
- biohazard bag

Procedure:

1. Use aseptic technique to remove a colony from the plate with a sterile inoculating loop.
2. Add the inoculum to the opened vial with beads and glycerol/broth.

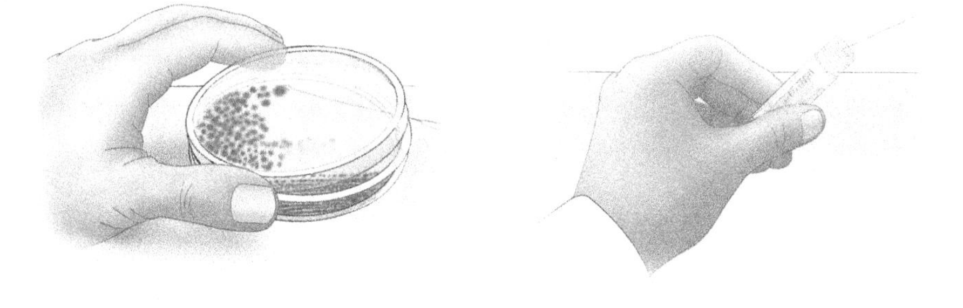

3. Screw on the cap. Mix the contents by vigorously shaking the vial. Do not use a vortex.

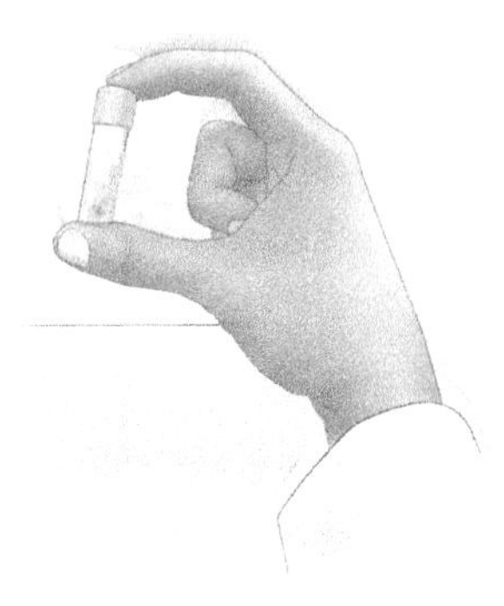

4. Remove the cap but do not set it down. Use sterile technique to remove all of the liquid in the vial with the Pasteur pipette. Place the excess fluid in a beaker with disinfectant. Discard the pipette in a biohazard bag.

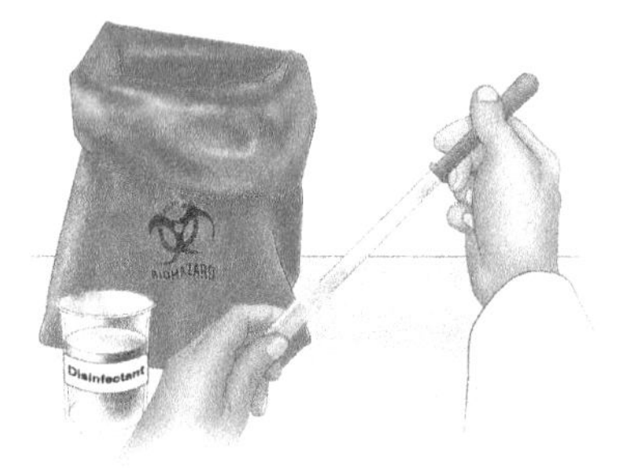

5. Screw on the cap so that it is finger tight. Store the vial at –70° C.

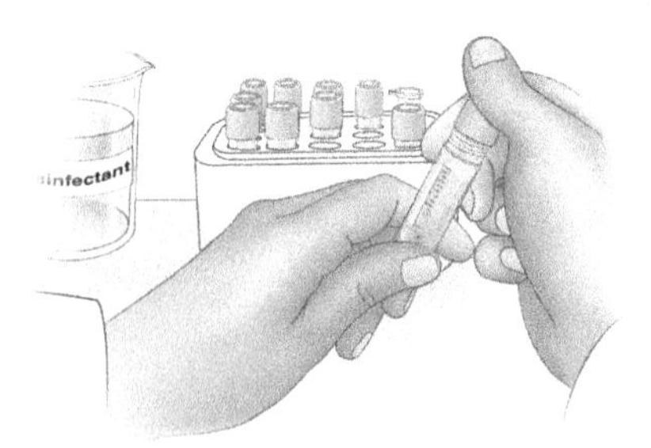

6. To recover frozen cells from the vial, remove a bead using a sterile forceps or a sterile inoculating needle bent into an L-shaped hook.

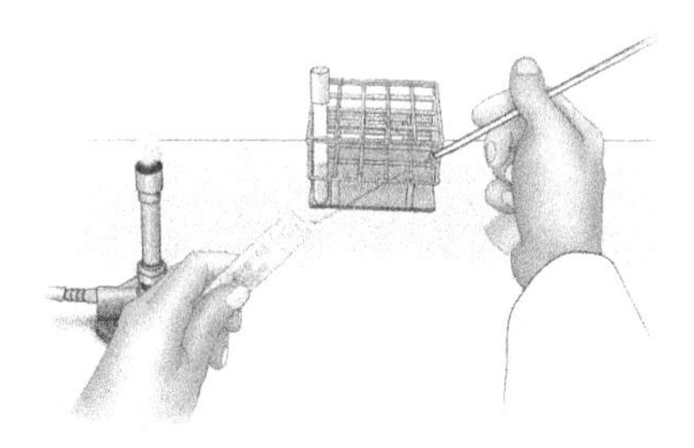

7. Use aseptic technique to place the bead in a tube with broth medium. Incubate the tube with the bead overnight at 35° C.

8. Screw the lid back on the vial until it is finger tight. Return the vial to the freezer as quickly as possible to prevent thawing.

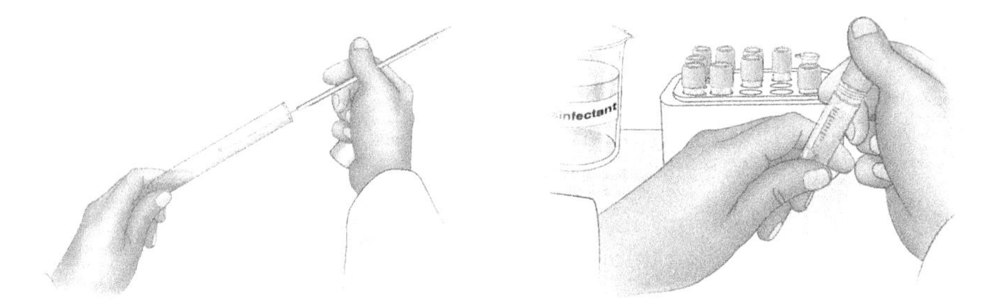

9. Flame the inoculating wire hook or forceps.

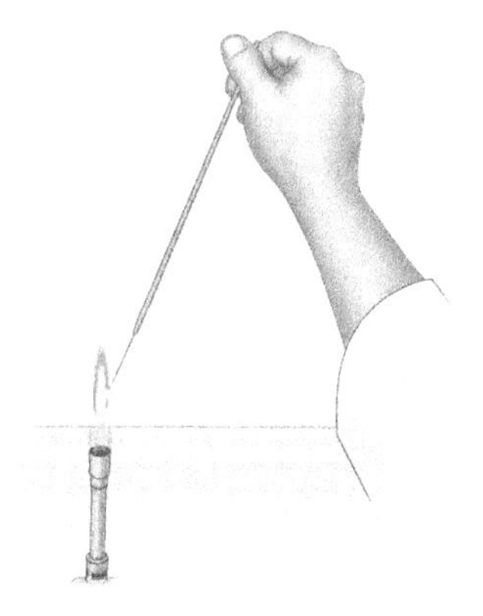

Troubleshooting

- For a reserve stock culture, prepare an agar slant from the broth tube after the revived bacteria have grown. Cell viability is much longer on an agar slant.
- Sterile glycerol that has been preheated in a 50° C water bath will be less viscous and, therefore, easier to pipette.
- If you store cryoprotected bacteria in a kitchen-type freezer (–20° C), the freezer must not be a self-defrosting model. The freeze–thaw cycle will drastically reduce viability of stored bacteria.
- Keep the time that a vial or tube containing glycerol/bacteria is out of the freezer to a minimum to prevent thawing.

CULTURING ANAEROBIC BACTERIA

Purpose:

- To cultivate bacteria that cannot survive in the presence of oxygen.

Strict (or **obligate**) **anaerobes** are bacteria that grow only in the absence of oxygen. These bacteria are apparently unable to detoxify reactive forms of oxygen created during metabolism.

Thioglycollate Broth

One laboratory technique for culturing anaerobes uses a broth supplemented with the reducing agent **thioglycollate**, which reacts with oxygen to form water. An indicator dye changes color where oxygen has dissolved in the broth (pink or blue: oxygen present; colorless: oxygen absent).

Any specimen that will be cultured for anaerobic bacteria must be collected under conditions that exclude or, at least, minimize exposure to oxygen.

Materials:

- thioglycollate broth tube
- bacterial culture in broth or on agar surface

Procedure:

1. Follow the instructions as described on pp. 29–31 for transferring an inoculum to a broth culture.

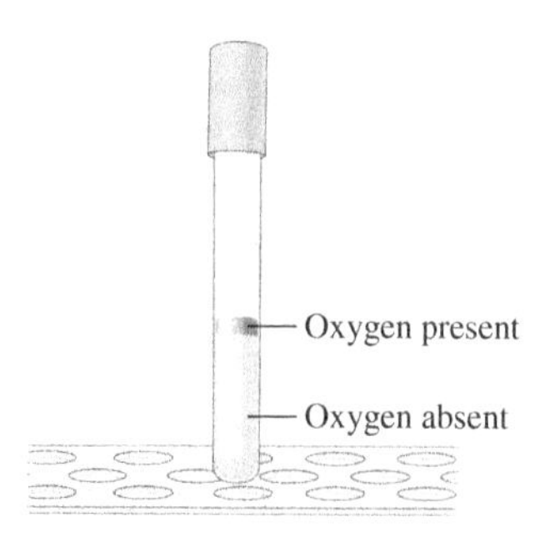

2. After 48 hours incubation, examine the tube for the location of bacterial growth.

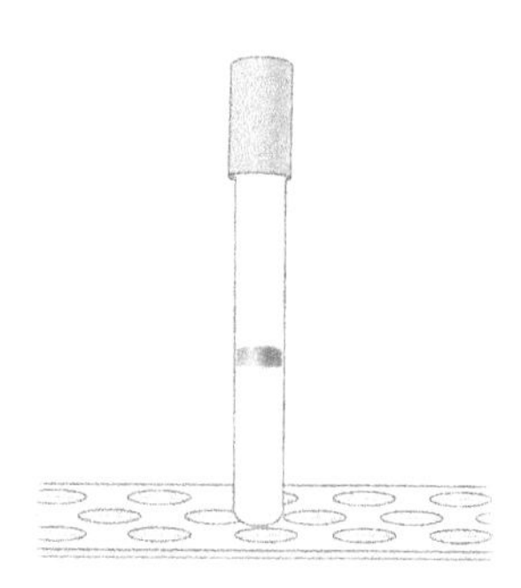

GasPak Anaerobic System

The **GasPak Anaerobic System** uses a chemical reaction to consume free oxygen gas in a sealed jar. Chemicals in a disposable envelope are activated upon addition of water. Hydrogen is generated and reacts in a catalyzed reaction with oxygen in the sealed jar to form water. Carbon dioxide, which stimulates the growth of some anaerobes, is another product of the reaction. An indicator strip inside the jar is colorless when oxygen is absent, blue when oxygen is present.

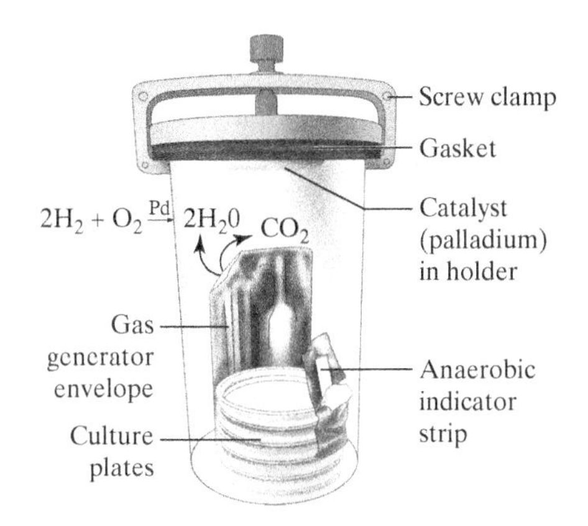

Materials:

- anaerobic jar with lid and screw clamp
- GasPak envelope
- water, distilled, sterile
- pipette, 10 mL
- pipettor
- GasPak anaerobic indicator strip in a packet
- agar plates, inoculated with bacteria

Procedure:

1. Place the inoculated plates in the anaerobic jar with the agar side up.

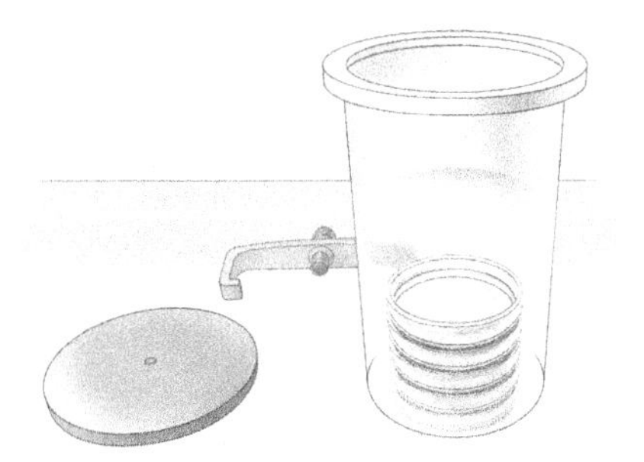

2. Cut off one corner of the GasPak envelope. Place the envelope on one side of the jar. Add 10 mL of sterile distilled water to the envelope.

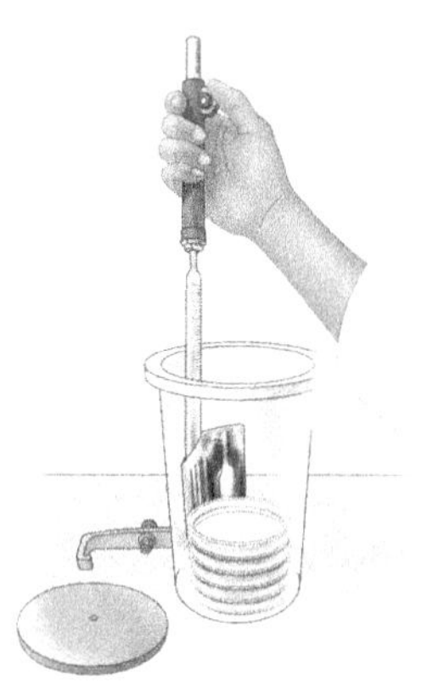

3. Open the packet with the indicator strip. Position the strip in the jar so it is visible.

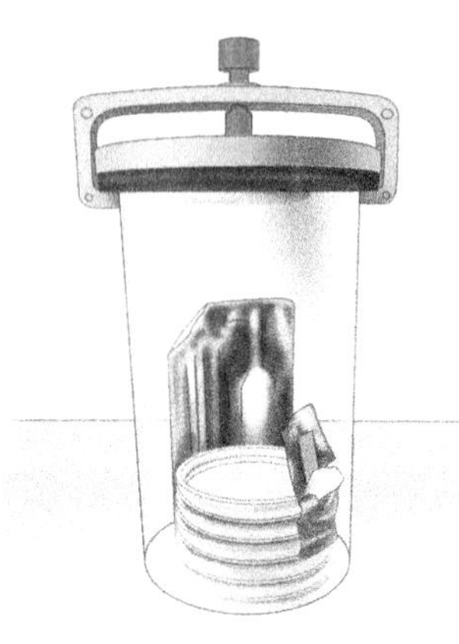

4. Secure the lid on the jar with the screw clamp. Tighten the clamp with your fingers only.

5. Place the jar in the incubator at 35° C for 24–48 hours. Check the color of the indicator strip after two hours; it should have no color if oxygen has been consumed.

Candle Jar

Some bacteria grow best when they are incubated in the presence of a carbon dioxide-enriched atmosphere. This environment can be created in a **candle jar**. After inoculated plates are placed in a jar, a scentless candle is lit inside the jar and the lid is screwed on tightly. The burning candle will generate sufficient levels of carbon dioxide in the jar (~5%).

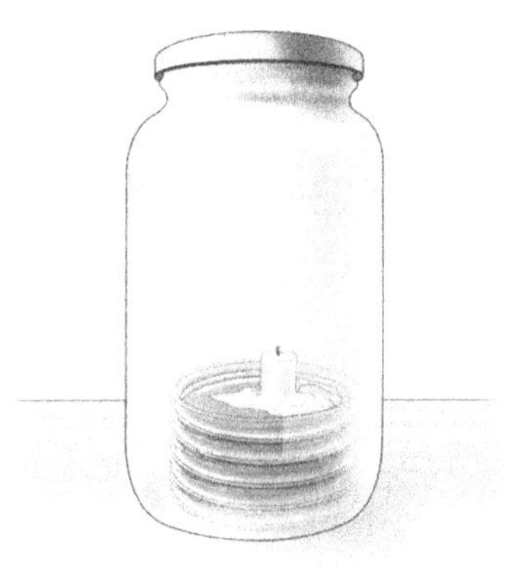

- *Do not have an open flame nearby when you add water to a GasPak envelope. Hydrogen gas is flammable. If hydrogen and oxygen are ignited in a closed space, they will explode.*

Troubleshooting

- If the color in a sterile thioglycollate broth tube extends through one-quarter of the volume, heat the tube to boiling for 1 minute to drive out dissolved oxygen. Allow the tube to cool before you inoculate it.
- If the color of the GasPak indicator strip in the sealed jar changes to blue, you must open the jar, remove the GasPak envelope, and activate another envelope.
- If no moisture appears on the inside walls of the sealed jar, the oxygen-removing reaction did not take place. You must begin again.

Effective Use and Responsible Care of the Light Microscope

Because the light microscope is an essential and expensive instrument in the microbiology lab, it is critical that every student uses it properly and cares for it responsibly. It must be kept clean and in working condition. Different models of microscopes are available in teaching laboratories. The following are guidelines; your instructor may provide additional instructions for the particular microscope(s) in your lab. While the procedures described here are for binocular microscopes, use of monocular microscopes is the same except for the number of ocular lenses.

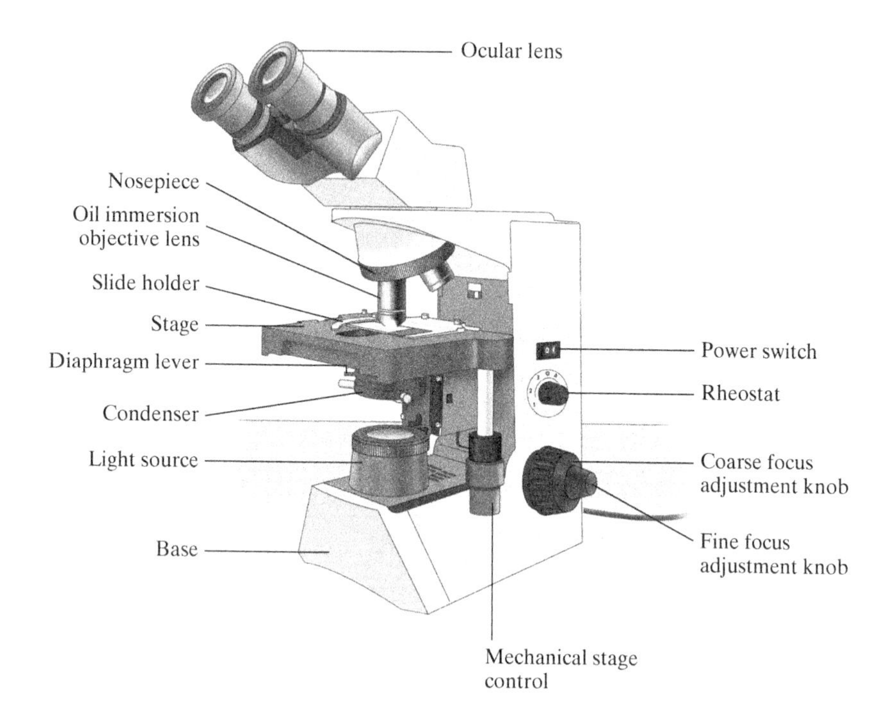

Materials:

- prepared microscope slides (wet mount, stained smear)
- microscope
- immersion oil
- lint-free lens paper

Procedure:

1. Use both hands to carry the microscope, one hand holding the arm and the other supporting the base.

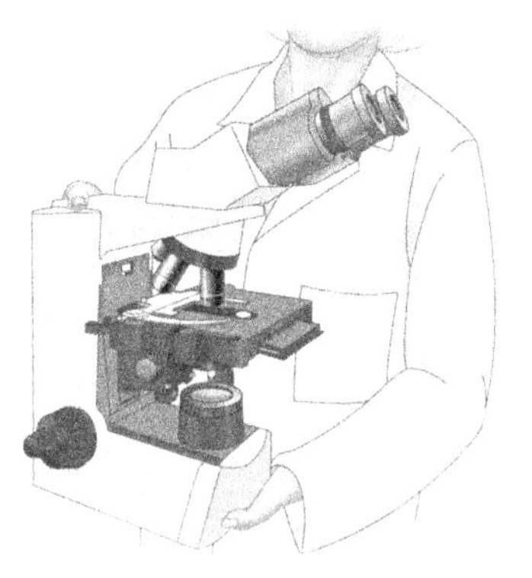

2. Position the microscope on the lab bench top away from the edge. Adjust the height of your seat so you can view through the ocular lenses comfortably.

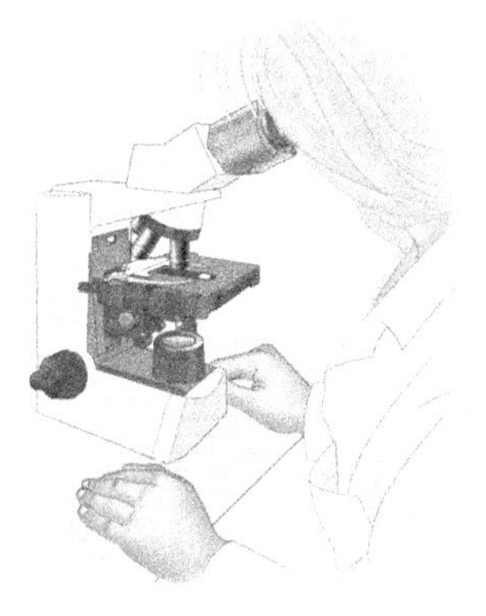

3. Clean the ocular and objective lenses with lens paper only.

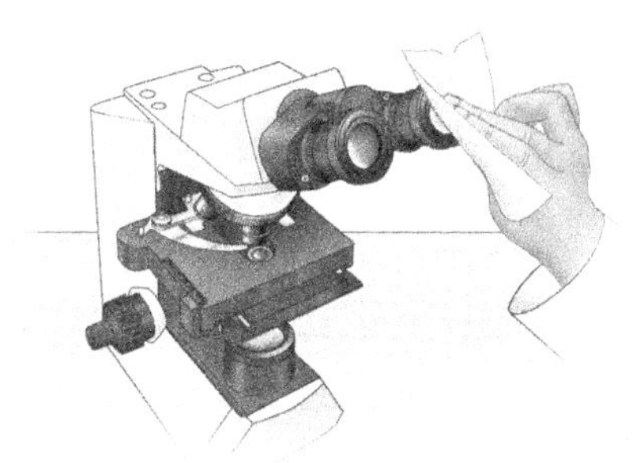

4. Hold on to the plug, not the cord, when you connect it to the electrical outlet.

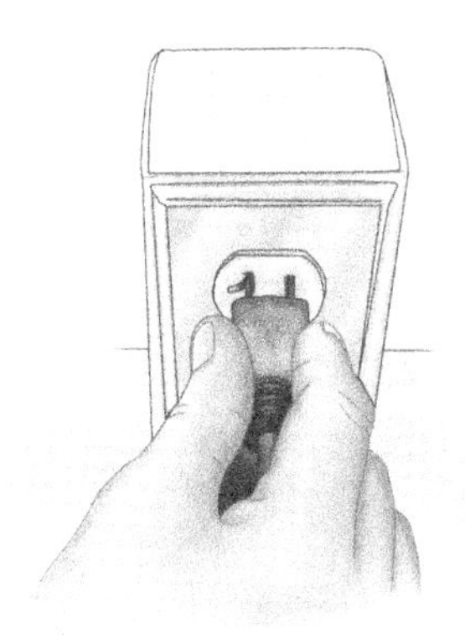

5. Turn on the power to turn on the light. Adjust the light intensity by slowly turning the rheostat knob.

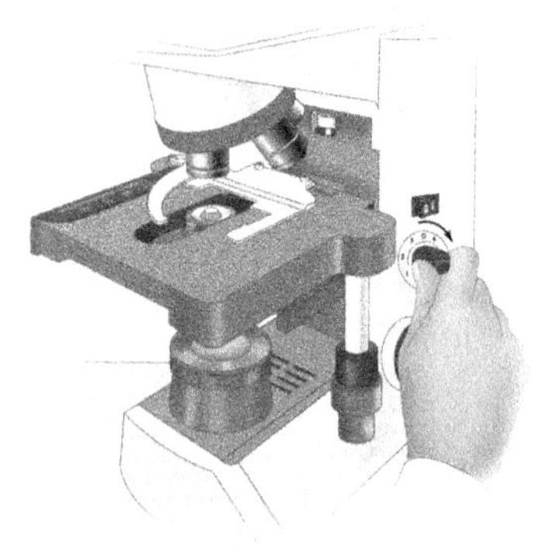

6. Raise the condenser to its maximum height.

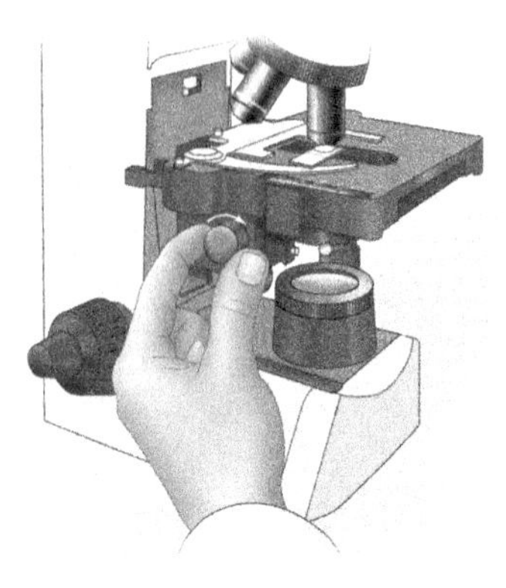

7. Place the slide with the specimen side up. Position the slide so that it is held snugly by the slide holder.

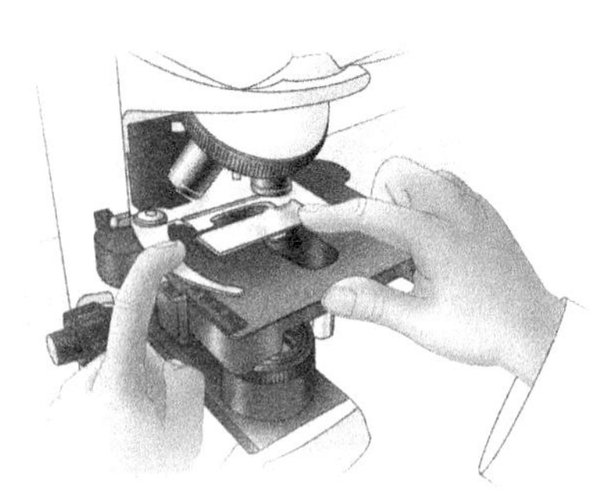

8. Adjust the distance between the ocular lenses so that you see only one image.

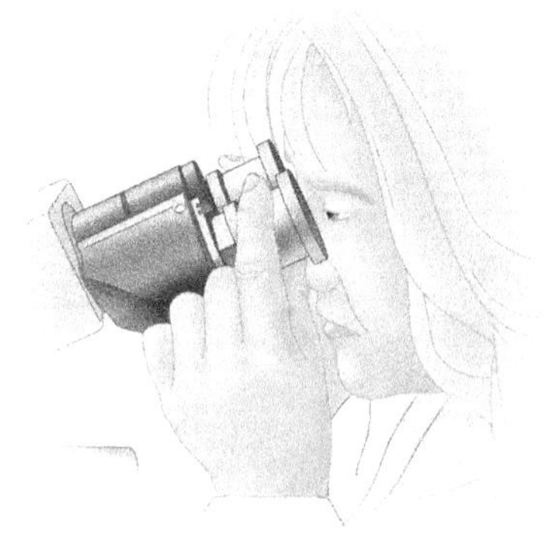

9. As you look on from one side, rotate the nosepiece to bring the 10X objective lens in place. You will feel a click as the lens locks into place.

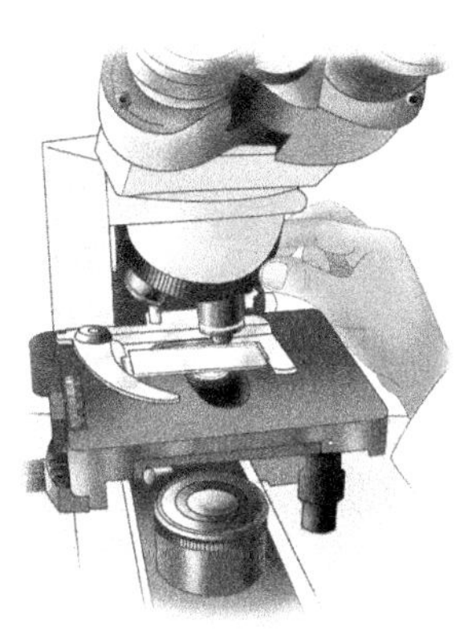

10. Center the slide over the light hole by turning the mechanical stage adjustment knobs. One knob moves the slide from side to side, the other forward and backward.

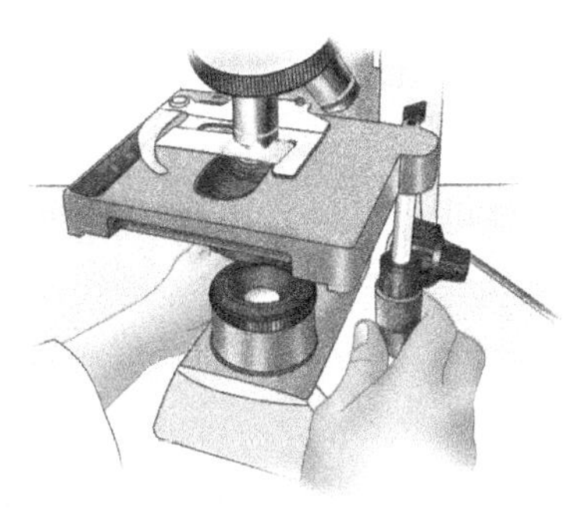

11. As you look on from one side, turn the coarse focus adjustment knob until the stage and objective lens are as close as possible. Note the direction in which you turn the knob.

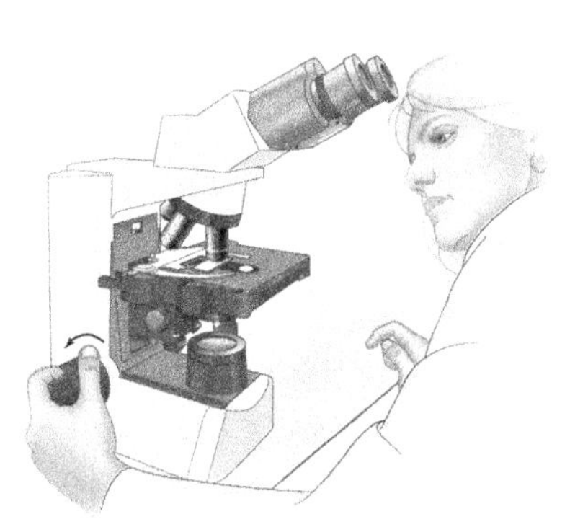

12. Slowly turn the coarse focus adjustment knob in the direction opposite from that in Step 11. The distance between the stage and the objective lens will increase.

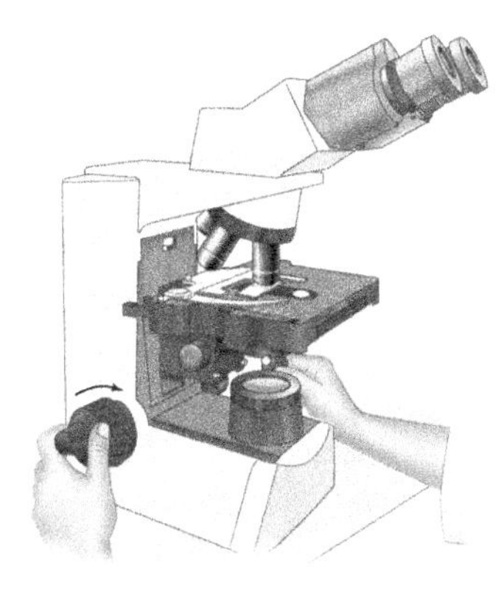

13. Adjust the light intensity by moving the diaphragm lever until the image has the best contrast against the background.

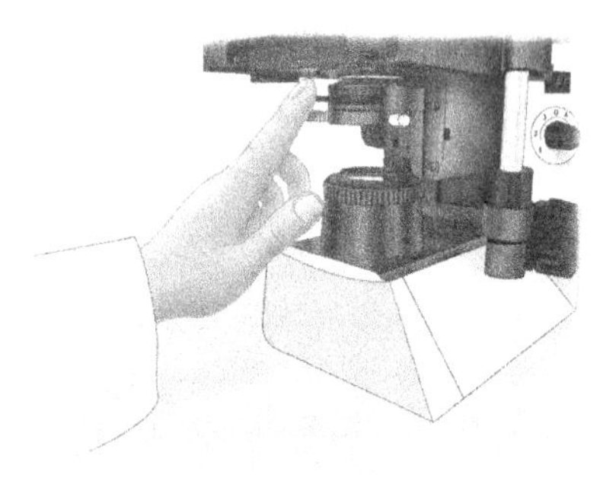

14. Slowly turn the fine focus adjustment knob to sharpen the focus. Do this with your right eye open and your left eye closed. If you turn this knob more than one complete turn, turn the knob back to its original position and repeat Steps 10–13.

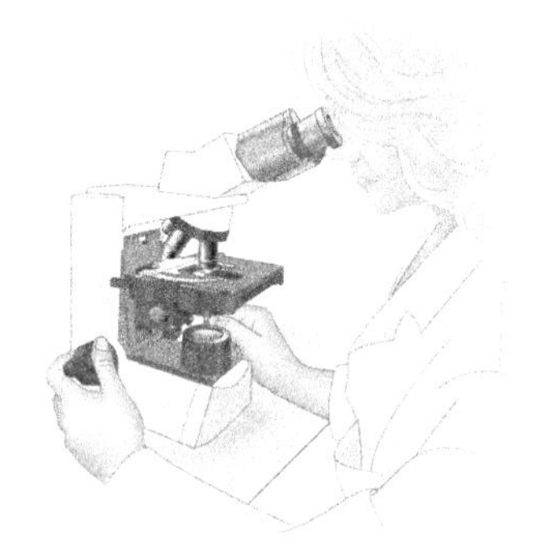

15. Open your left eye and close your right eye. Use the left eyepiece adjustment ring to focus for your left eye.

16. When you have finished your observations with the 10X objective, rotate the nosepiece and click the 40X objective into place. Because of the optics of the microscope, the specimen will be almost in focus already; you will only need to use the fine focus adjustment knob. You may need to increase the light intensity.

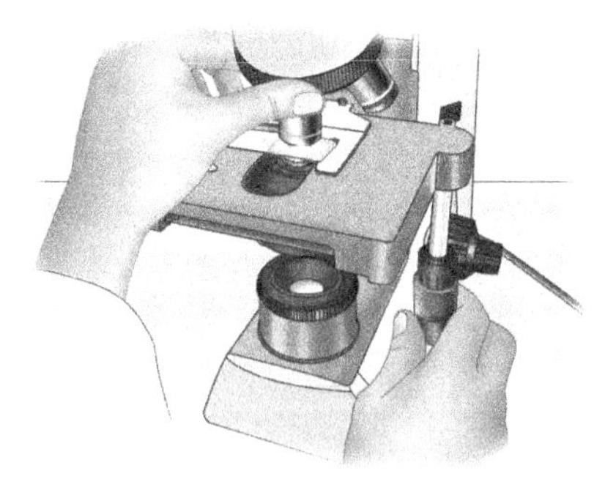

17. Use the oil immersion lens (100X) to view stained bacterial smears. Turn the nosepiece halfway between the 40X and 100X objectives, and place a drop of immersion oil on the place on the slide that you wish to view. The diaphragm should be fully opened.

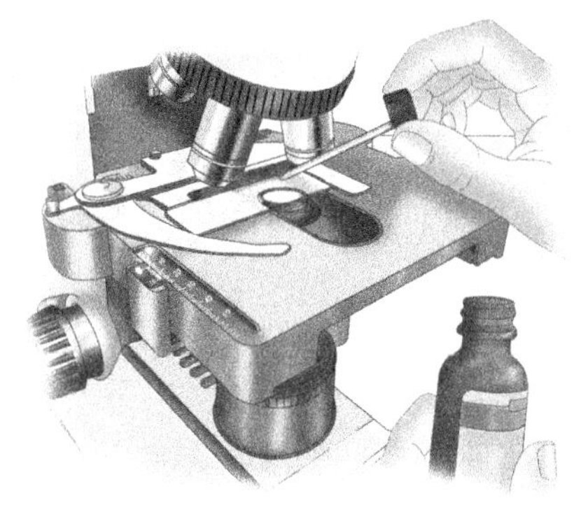

18. Turn the nosepiece to click the 100X objective into position in the oil. Turn the fine focus adjustment knob slowly to get a sharp image. You always need to move the objective lens and stage away from each other to prevent damage to the lens.

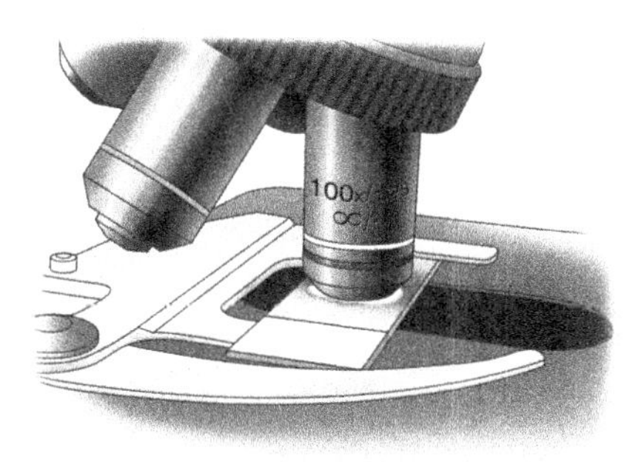

19. When you have finished your observations, turn the nosepiece halfway between the 100X and the lowest power objectives. Before you remove the slide, turn the coarse focus adjustment knob to increase the distance between the lens and stage. Use lens paper to remove the oil from the 100X objective.

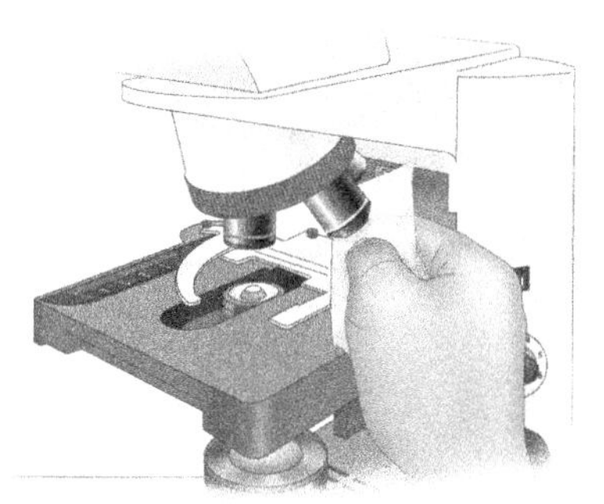

20. Turn off the light. Hold the plug to remove it from the outlet; do not pull it out by the cord. Wrap the cord around the base or holder.

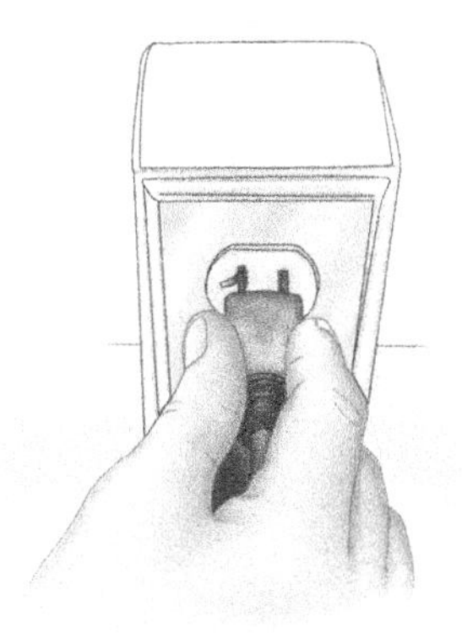

21. Turn the nosepiece to position the lowest power objective lens in place. Turn the coarse focus adjustment knob to bring the lens and stage as close as possible.

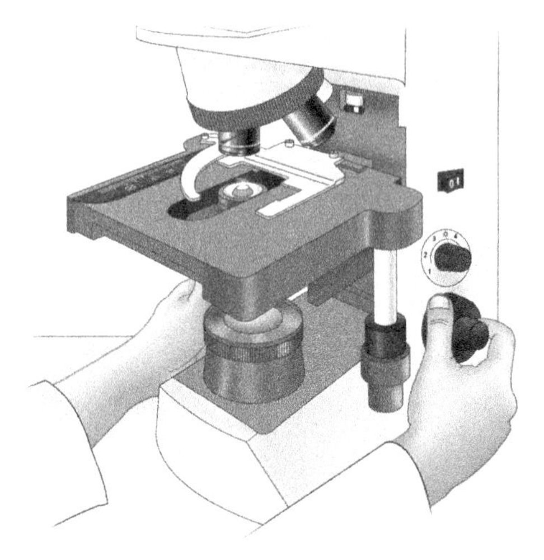

When you become more experienced with using the oil immersion lens for viewing stained bacteria, you can skip the lower magnifications and focus from the start with the 100X objective.

1. Position a slide with a stained smear in the center of the light hole.

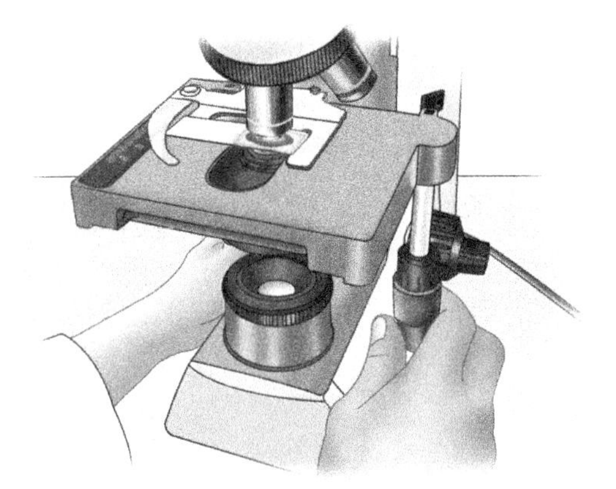

2. Place a drop of immersion oil on the smear.

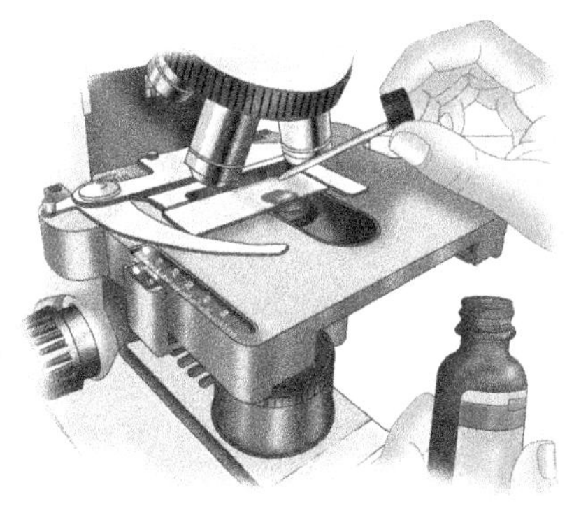

3. Turn the nosepiece to click the oil immersion lens into place.

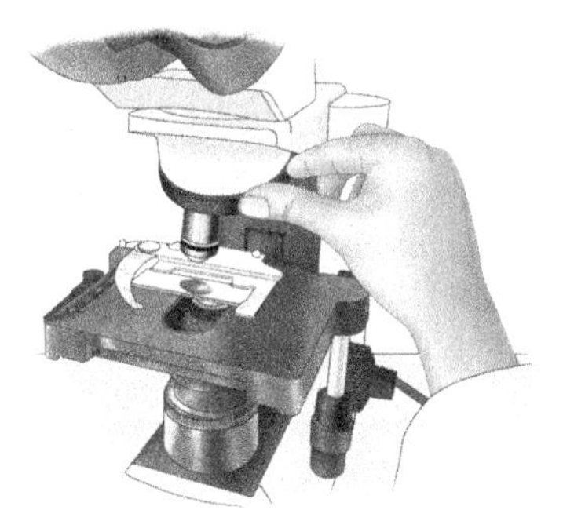

4. As you look on from one side, turn the coarse focus adjustment knob to bring the oil immersion lens into the oil and almost touching the slide surface. Note the direction in which you turn the knob.

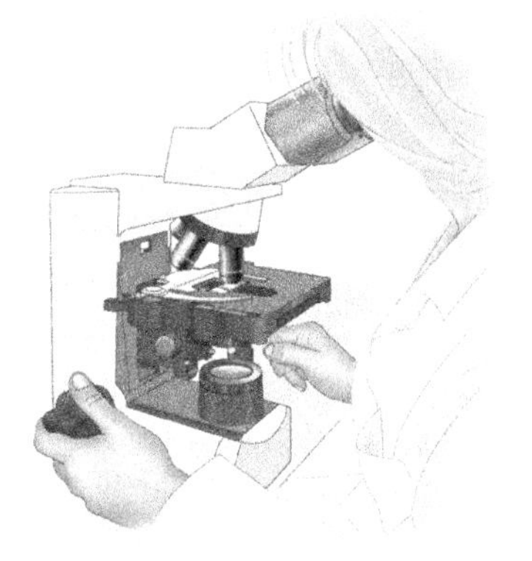

5. Then, while you view through the ocular lenses, very slowly turn the coarse focus adjustment knob in the direction opposite of Step 4. As soon as you see a "flash" of an image, use the fine focus adjustment knob to sharpen the image.

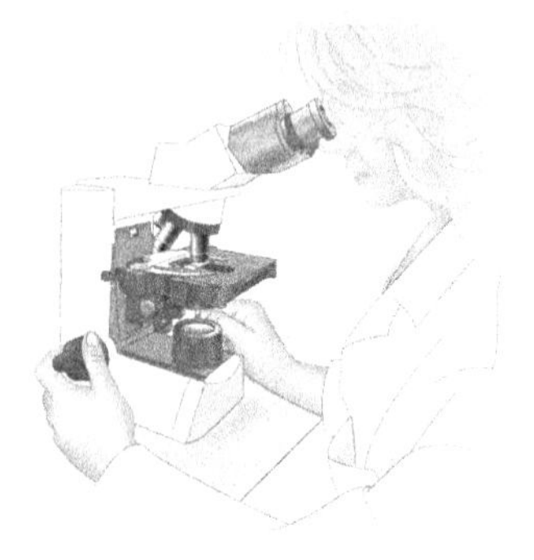

- *If you break a slide, take care that you do not cut yourself as you place the pieces in the glass disposal container.*
- *Because the distance between the oil immersion lens and the slide is only 0.1 mm, do not turn the coarse focus adjustment knob without watching from the side. Otherwise, you might force the lens into the slide, breaking the slide and perhaps damaging the lens.*
- *If you are viewing a live bacterial preparation, dispose of the slide in disinfectant.*
- *While unplugging, if you pull on the cord rather than holding the plug, the connection and insulation between the cord and plug will eventually loosen. This will expose bare wire carrying electricity, probably giving you a shocking experience.*

Troubleshooting

- If the light intensity is too high, the contrast between microbes and the background will suffer.
- As the magnification increases, the light intensity must be increased by enlarging the diaphragm opening and/or turning up the rheostat.
- Remove all the oil from the 100X objective with lint-free lens paper every time you use the lens and at the end of the lab period. Do not use Kimwipes, paper towels, bibulous paper, facial tissues, or toilet paper.
- Turn the nosepiece to the objective lens with the lowest power when you remove a slide viewed with the oil immersion lens.
- When you turn the nosepiece to change objective lenses, do not allow the 40X objective to touch immersion oil on the slide.

- You may not be successful in viewing stained bacteria magnified with the oil immersion lens for any of these reasons:
  - You did not apply enough oil over the specimen. This causes the bacteria to appear "fuzzy." Add more oil.
  - The oil has bubbles that mask the cells. Separate the slide from the lens by turning the coarse focus adjustment knob. Use lens paper to remove the oil from the lens. Check to see if the slide needs additional oil. Refocus.
  - The preparation has too few bacteria. Prepare a new slide.
  - You did not center the preparation before you switched to the higher magnification. Use the mechanical stage adjustment knobs to center the preparation.
  - The slide is upside-down on the stage.
  - The fine focus adjustment knob has been turned too far by previous users. Return it to the middle position.
  - You turned the fine focus adjustment knob so quickly that you missed the specimen image.
  - You turned the fine focus knob too far, so that the lens is no longer immersed in the oil. Be sure that when you reverse directions with the fine focus adjustment knob you return the lens to its original distance from the slide.
  - At 10X, you focused on stained "junk" located on the underside of the slide.
  - An ocular lens is dirty. Clean it with lens paper.
  - The light intensity is either too high or too low. Adjust with the diaphragm lever and/or the rheostat knob, or the slide control.

Viewing Live Microorganisms

Preparing a Hanging-drop Slide

Purpose:

- To observe live microbes for size, shape, and arrangement of cells.
- To observe live bacteria that use flagella for motility.

Materials needed:

- clean depression microscope slide (wipe well with lens paper)
- clean coverslip (wipe carefully with lens paper)
- petroleum jelly
- inoculating loop
- bacterial culture in broth

Procedure:

1. Apply a thin film of petroleum jelly on the side of your palm.

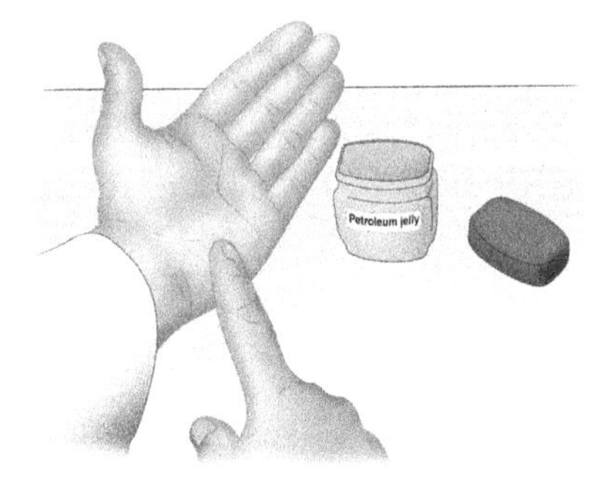

2. Hold the coverslip between your forefinger and thumb and gently rub the edge of the coverslip in the petroleum jelly. Do this on all four edges to form a thin ridge.

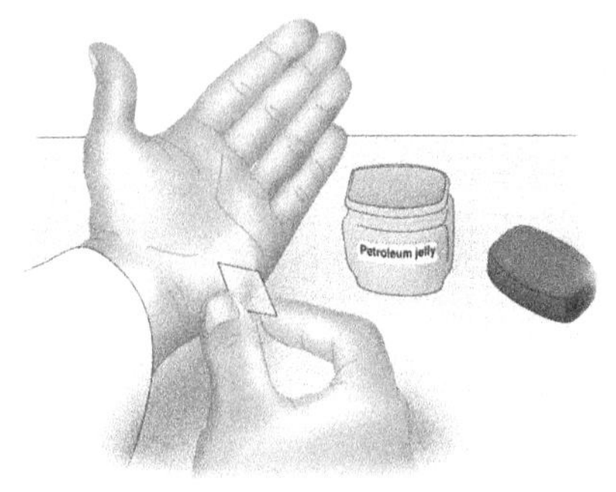

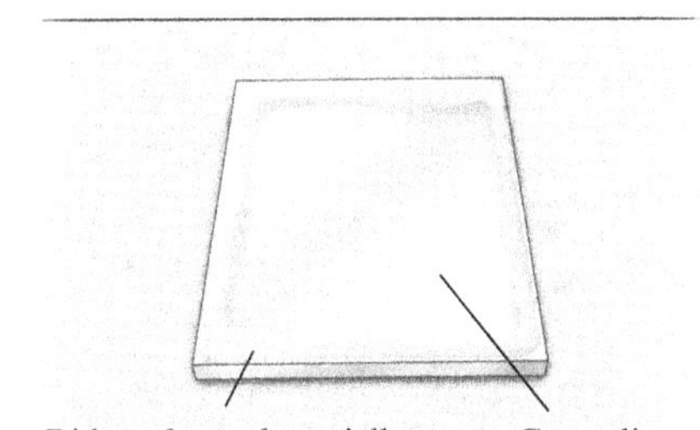

3. Place the coverslip jelly-side up on a paper towel on the lab bench top.

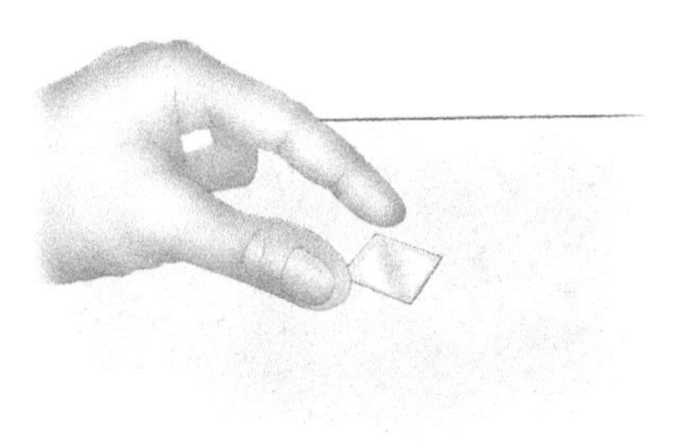

4. Use a flamed inoculating loop to transfer a loopful of bacteria to the center of the coverslip.

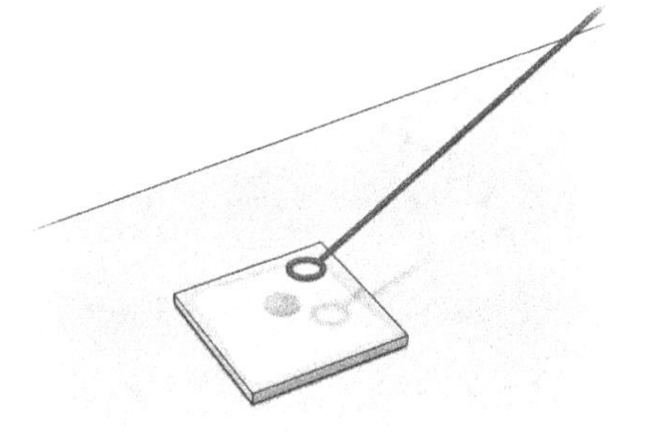

5. Flame the inoculating loop before you place it on the lab bench top.

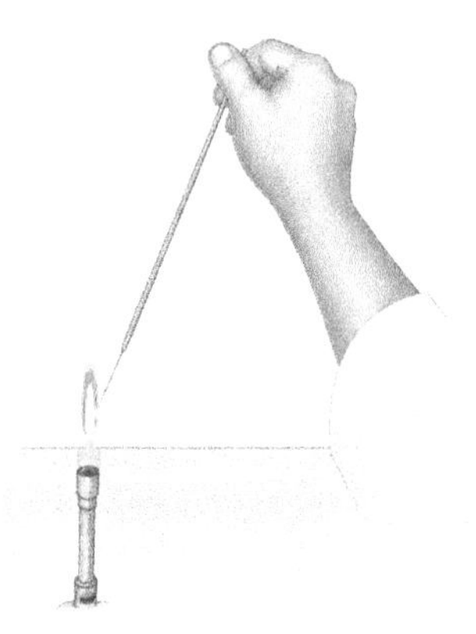

6. Center the cavity in the depression slide over the coverslip. Lower the slide until the petroleum jelly ridge on the coverslip creates a seal.

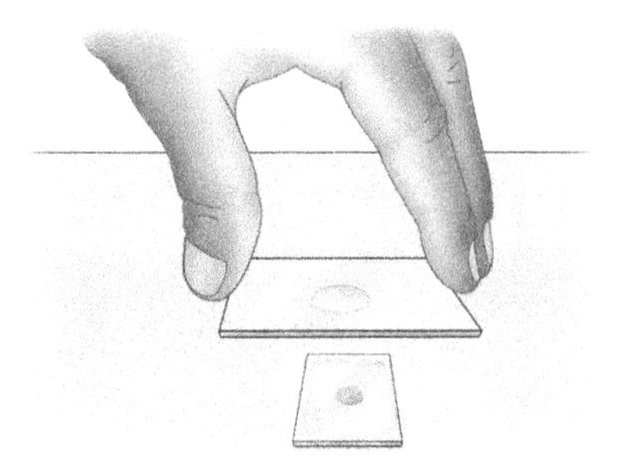

7. Quickly turn the slide over. The drop of bacteria is now suspended in the depression.

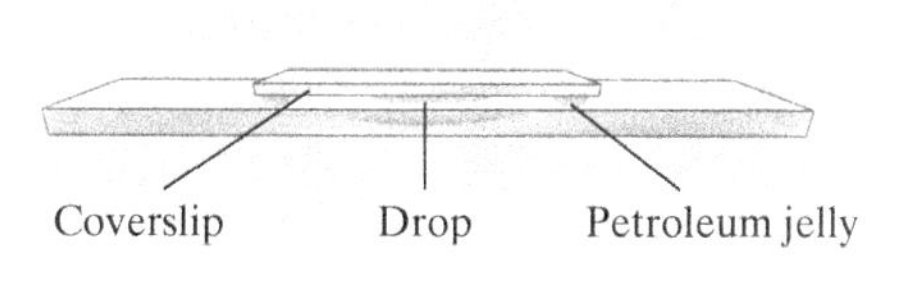

8. Examine the drop under low-power magnification (10X). Adjust the amount of light to a low level with the diaphragm lever. (Because unstained bacteria are transparent, minimum light is needed.)

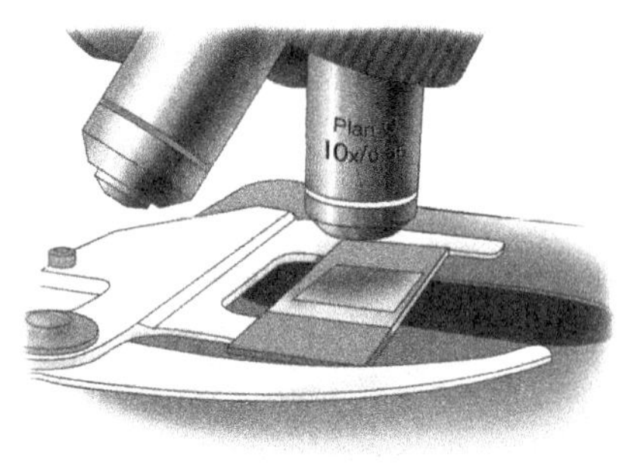

9. Examine the drop under high-dry magnification (40X). Increase the amount of illuminating light.

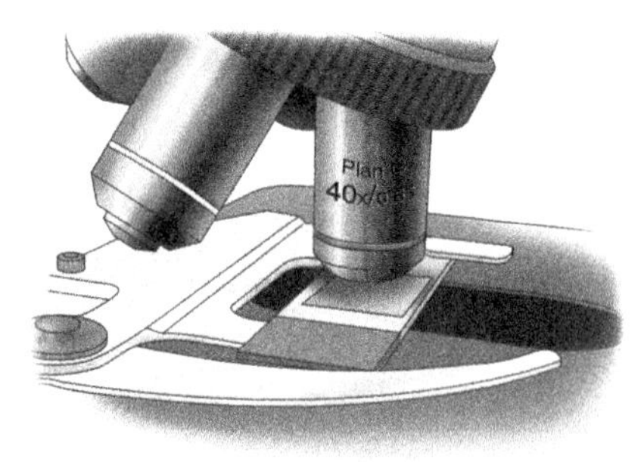

10. When you are finished, place the slide in disinfectant to kill the bacteria. The slide can also be autoclaved.

STAINING BACTERIA

Preparing a Bacterial Smear

Purpose:

- To prepare a thin film of bacteria spread on a microscope slide before staining the cells.
- To fix an air-dried film to prevent cells washing off during staining, and to denature bacterial enzymes that could digest the cell wall.

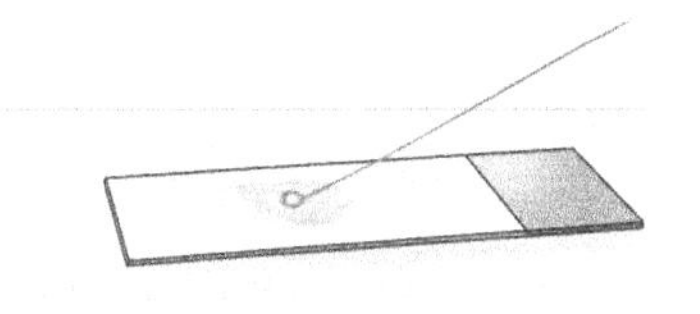

A bacterial smear is prepared by spreading the specimen into a thin film on a clean microscope slide

Materials needed:

- bacterial culture in a broth tube, agar slant tube, or on an agar plate
- clean glass microscope slide (1 × 3 in.), preferably with a frosted end
- inoculating loop

Procedure:

1. Write the name of the bacterium at one end of the slide.

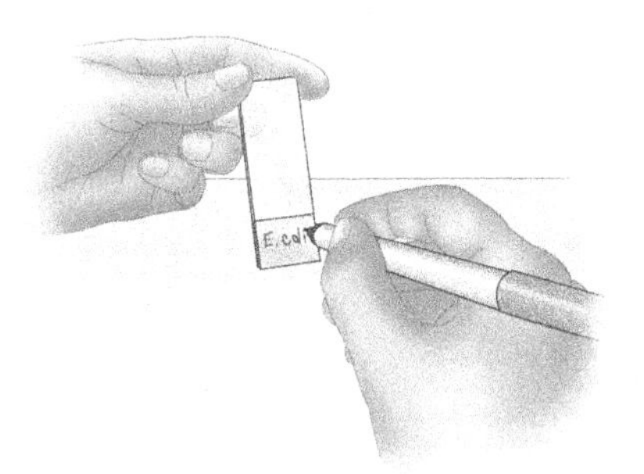

Preparing a smear

from a culture on a solid agar surface

1. Place a loopful of distilled water in the center of the slide.

from a culture in a liquid medium

1. Transfer a loopful of liquid culture to the center of the slide using a sterile loop.

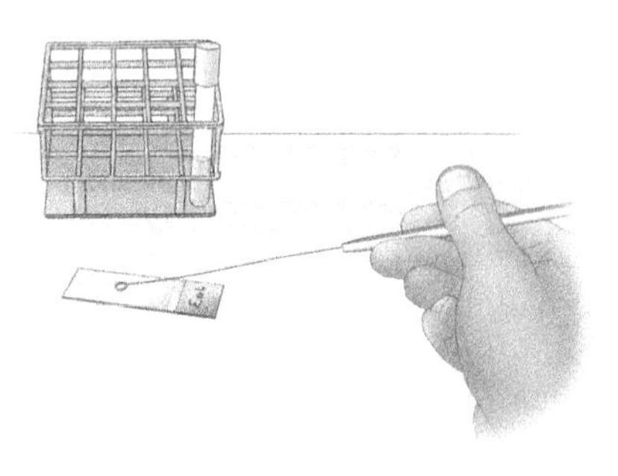

from a culture on a solid agar surface

2. Transfer a small amount of culture from the agar surface into the water drop using a sterile loop. Spread the mixture into a thin film.

from a culture in a liquid medium

2. Spread the drop of culture into a thin film.

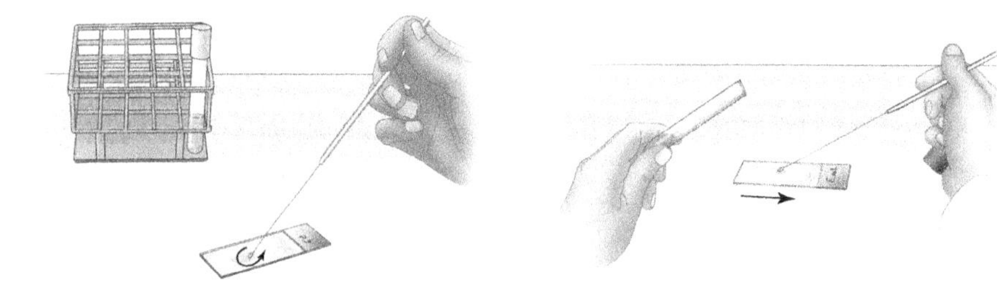

3. Flame the loop before putting it down.

4. Allow the smear to air dry.

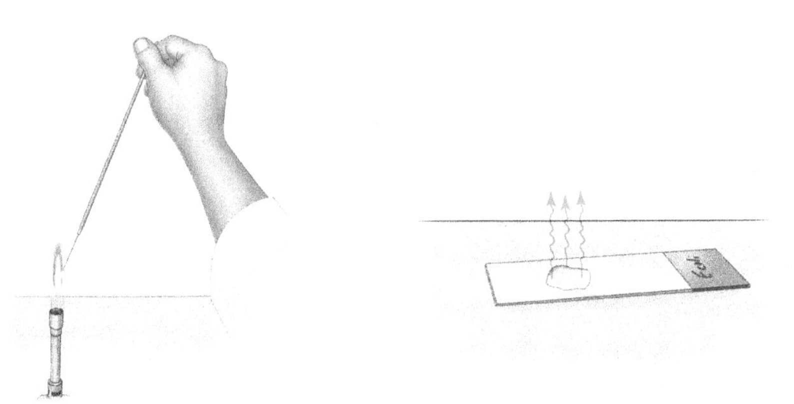

Use one of three methods to fix the dried smear:

5a. Pass the slide quickly through flame three times, smear side up. The smear is now said to be heat-fixed.

5b. Place the slide on a slide warmer (60° C for 10 minutes).

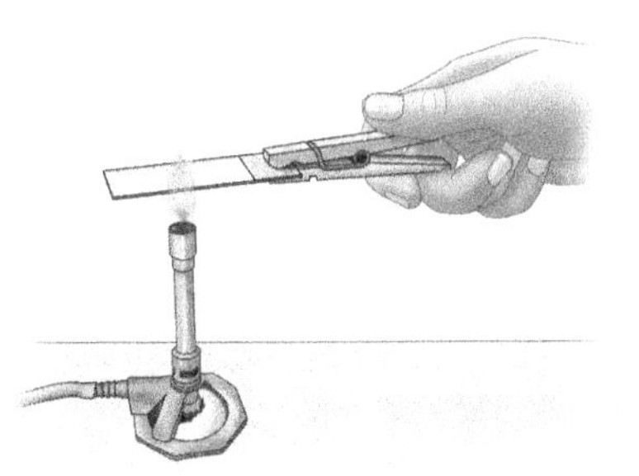

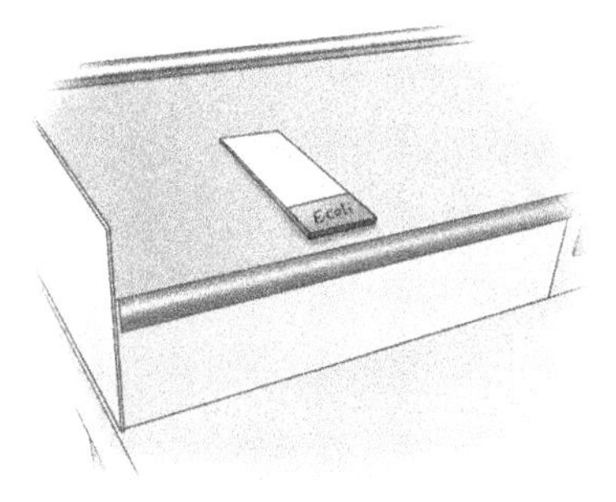

5c. Add 95% methanol to the dried smear for at least 1 minute.

Rinse off the methanol.

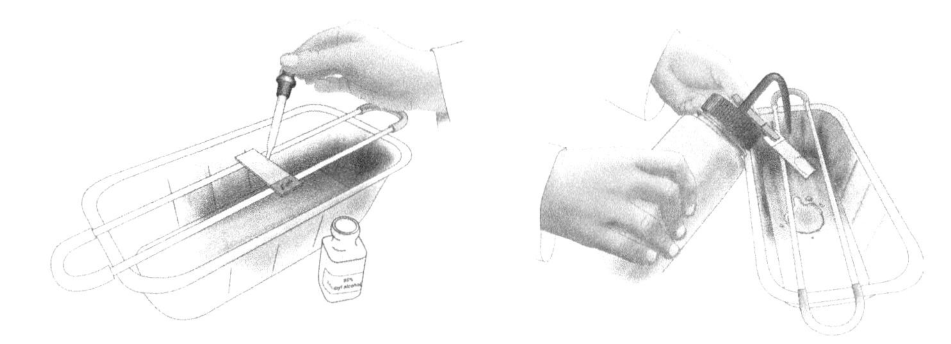

Preparing a smear from an environmental or clinical sample on a cotton swab

1. Roll the cotton swab in the center of the slide.

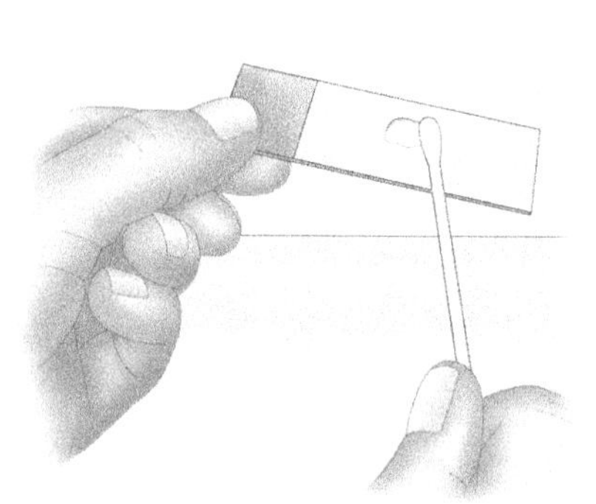

2. Discard the swab in the proper receptacle for sterilization.

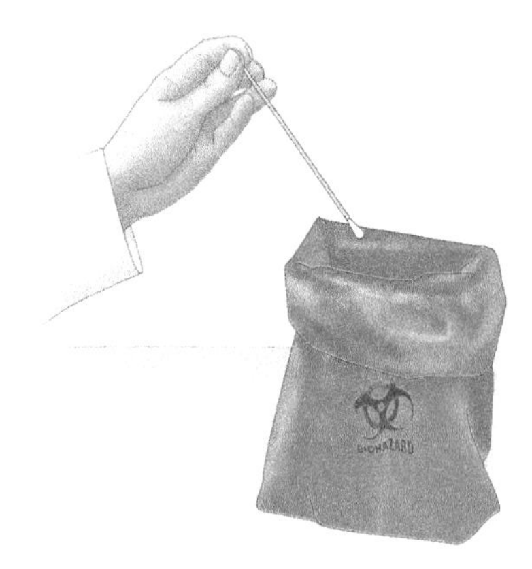

3. Allow the smear to air dry.

4. Fix the smear by heating or with methanol, as described above.

- Position open flames so that you will not burn yourself. Tie back long hair back to prevent it catching on fire.
- Allow a flamed inoculating loop to cool before you remove bacteria from a culture. Spattering caused by a loop that is too hot can create an aerosol.
- Do not set a used inoculating loop down without sterilizing it—it will contaminate the work surface.
- Hold the microscope slide with a clothespin or forceps when you fix a smear by flaming in a burner.
- Take care not to cut yourself if a glass slide breaks, especially during heat-fixing with an open flame.
- Dispose of broken glass in the designated container.
- If methanol gets on your skin, wash it off immediately.

Troubleshooting

- Placing bacteria in the center of the slide will make it easier to locate the stained smear under the microscope.
- Thin smears are necessary:
  - crowding of cells in a thick smear will make it difficult to observe cell shape, arrangement, and size.
  - thick smears may not be completely decolorized in a differential stain, such as the Gram stain.
- A sufficient number of bacteria must be placed on the slide to maximize detection of stained cells.
- If the smear is heated too much during fixation with an open flame, shapes of bacterial cells may become distorted.

Staining Bacteria
Preparing a Simple Stain

Purpose:

- To stain bacteria so they can be seen more clearly under the microscope.
- To determine the shape, arrangement, and size of bacteria in a fixed smear.

Materials needed:

- slide with fixed smear of bacteria
- stains in bottles with droppers:
  - crystal violet
  - methylene blue
  - safranin (red)
  - carbolfuchsin (red)
- staining rack over a tray or sink
- wash bottle with distilled or deionized water

  or

  rubber tubing connected to a faucet
- bibulous paper
- clothespin or forceps

Procedure:

1. Cover the fixed smear with several drops of stain for the times indicated below.

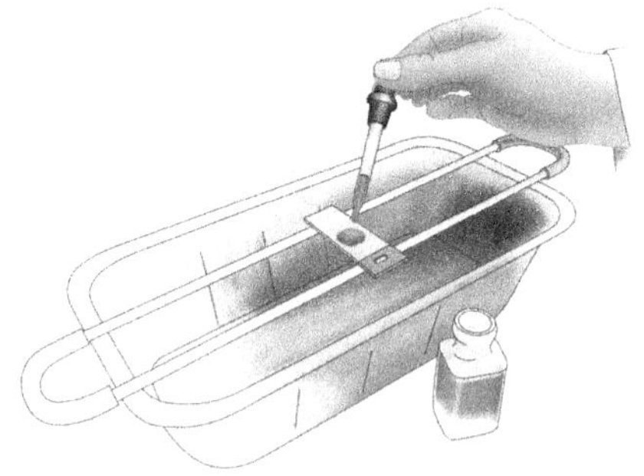

Staining times

carbolfuchsin	5–10 seconds
crystal violet	20–30 seconds
methylene blue	at least 1 minute
safranin	at least 1 minute

2. Rinse the slide with water to remove excess stain.

3. Blot water from the slide with pieces of bibulous paper.

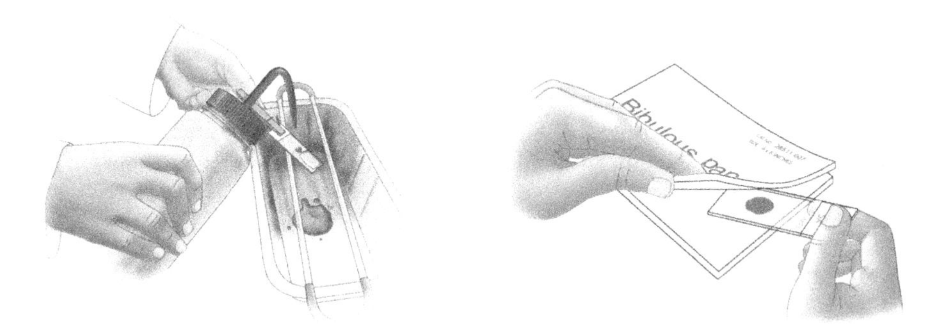

4. Examine the stained smear under the microscope using the oil immersion lens.

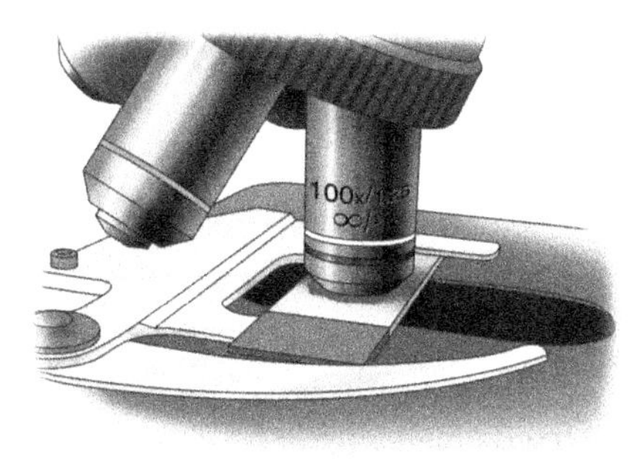

- *Take care not to cut yourself if a glass slide breaks, especially when blotting the slide dry. Dispose of broken glass in the designated container.*
- *Be careful not to get stain on your clothing—it does not wash out. Clean stain from your fingers with stain-removing cream.*

Troubleshooting

- If you blot the slide too vigorously with bibulous paper, you may rub off the stained smear.

Staining Bacteria

Preparing a Gram Stain

Purpose:

- To identify most bacteria by dividing them into two groups: gram-positive and gram-negative.
- To observe bacteria under a microscope for cell shape, arrangement, and size.

Materials needed:

- slide with fixed smear of bacteria
- stains in bottles with droppers:
  - crystal violet
  - Gram's iodine
  - 95% ethanol (*or* 1:1 95% ethanol:acetone)
  - safranin
- staining rack over a tray or sink
- wash bottle with distilled or deionized water
  or
  rubber tubing connected to a faucet
- bibulous paper
- clothespin or forceps

Procedure:

1. Place a slide with a fixed smear on a rack over a staining tray or a sink.

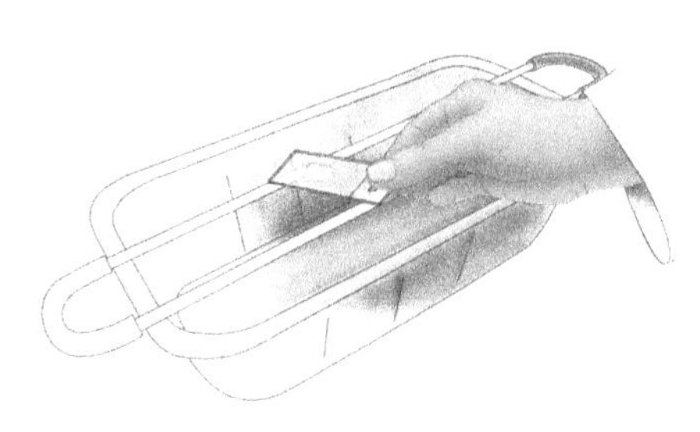

2. Cover the smear with crystal violet for 20 seconds.

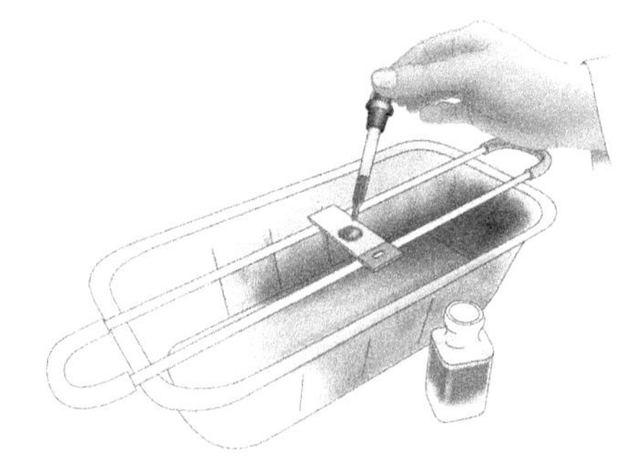

3. Rinse the slide with water to remove excess crystal violet.

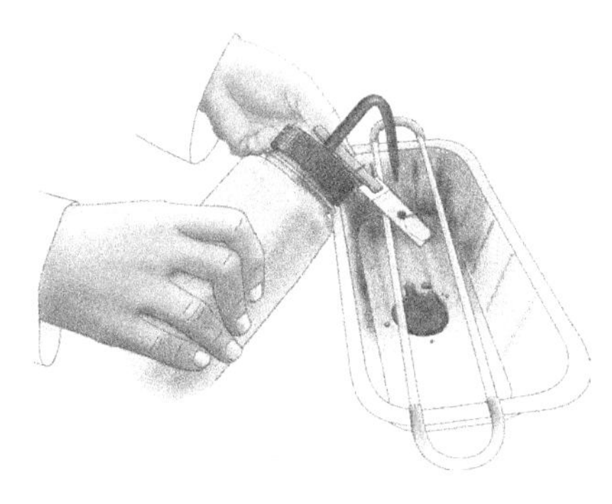

4. Cover the smear with Gram's iodine for 1 minute.

5. Rinse the slide with water to remove excess iodine solution.

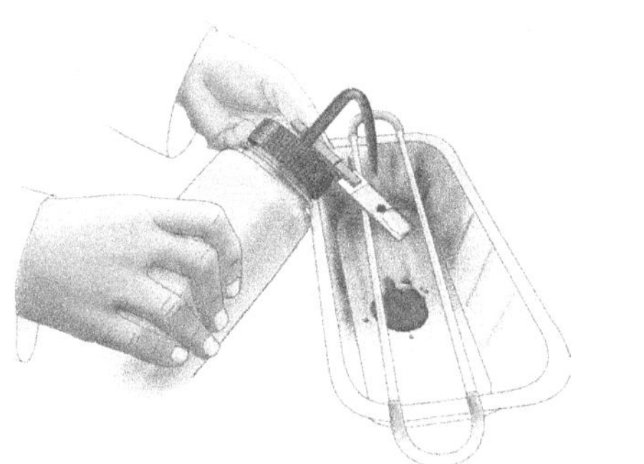

6. Decolorize with 95% ethanol or ethanol/acetone. Hold slide at a 45° angle while adding decolorizing reagent drop by drop until color stops running.

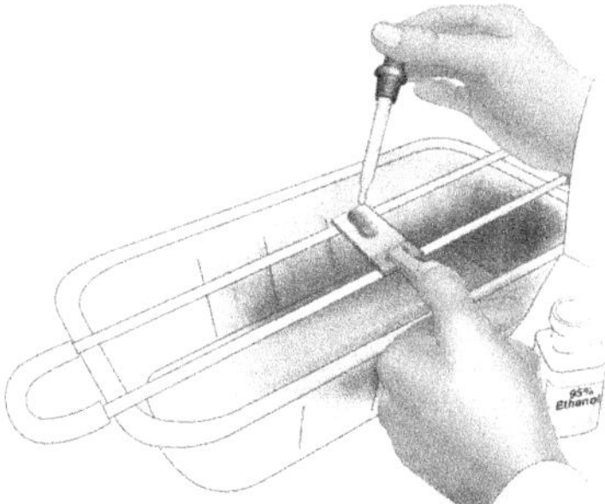

7. Immediately rinse the slide to remove the decolorizing agent.

8. Cover the smear with safranin for 1 minute.

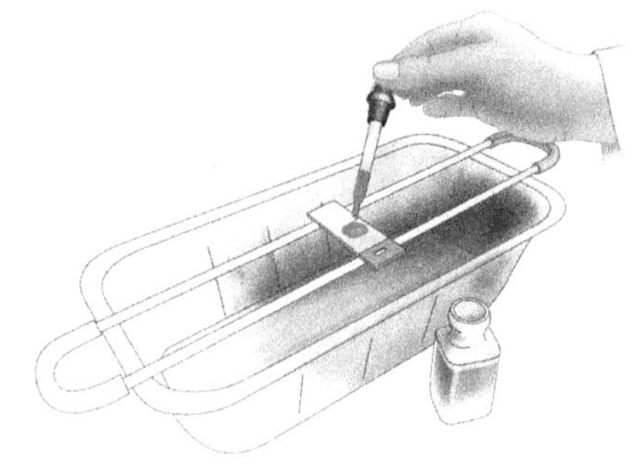

9. Rinse the slide with water to remove excess safranin.

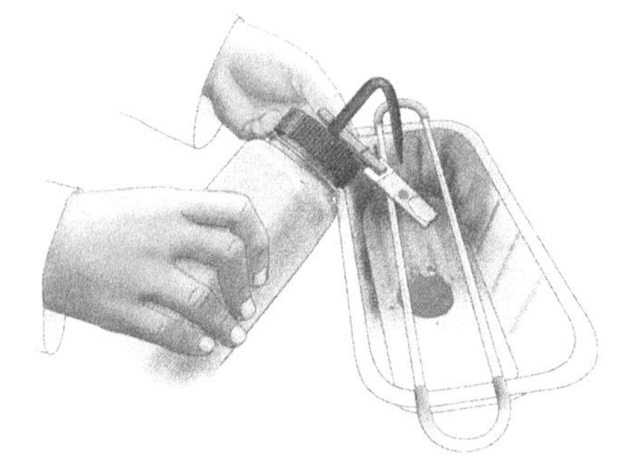

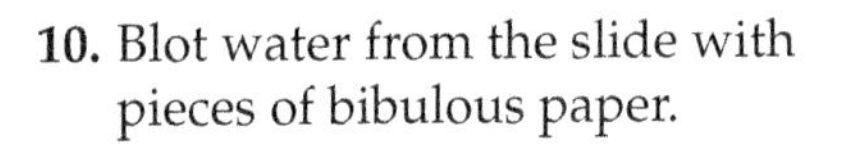

10. Blot water from the slide with pieces of bibulous paper.

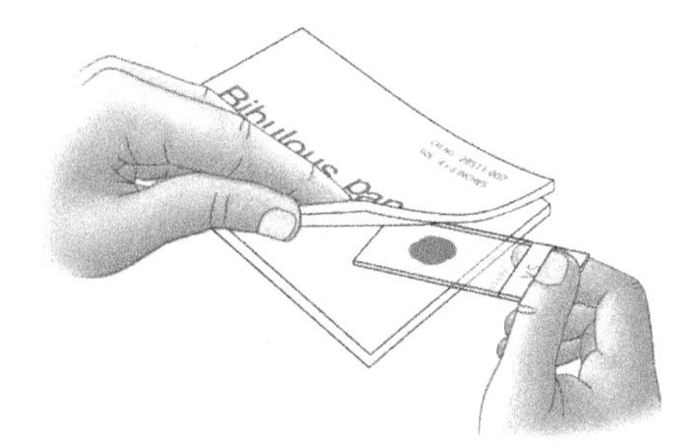

11. Examine the stained smear under the microscope using the oil immersion lens.

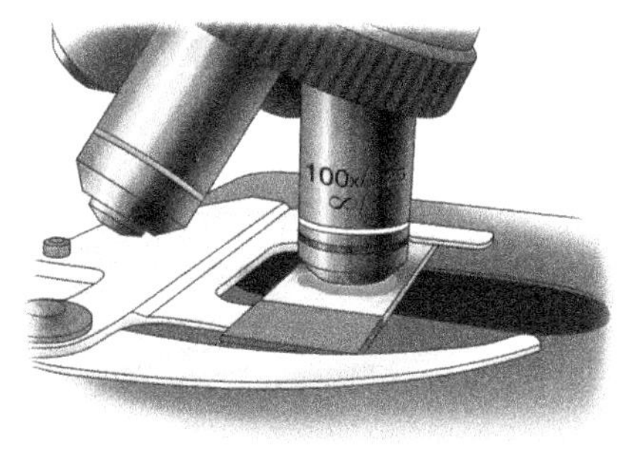

- *Keep the bottle and dropper of alcohol away from an open flame—alcohol is flammable.*
- *Take care not to cut yourself if a glass slide breaks, especially when blotting the slide dry. Dispose of broken glass in the designated container.*
- *Be careful not to get stain on your clothing—it does not wash out. Clean stain from your fingers with stain-removing cream.*

Troubleshooting

- If the culture used to prepare the smear is older than 18 hours, some gram-positive bacteria will appear gram-negative. Cell wall changes in older gram-positive cells can give inaccurate gram-negative or gram-variable results.
- If the Gram's iodine is pale yellow, it has aged and lost its potency.
- If too much decolorizing agent is applied, too much of the crystal violet/iodine complex will wash out from gram-positive cell walls.
- If you blot the slide too vigorously with bibulous paper, you may rub off the stained smear.

Appearance of bacterial cells in a fixed smear at each step in the Gram stain procedure

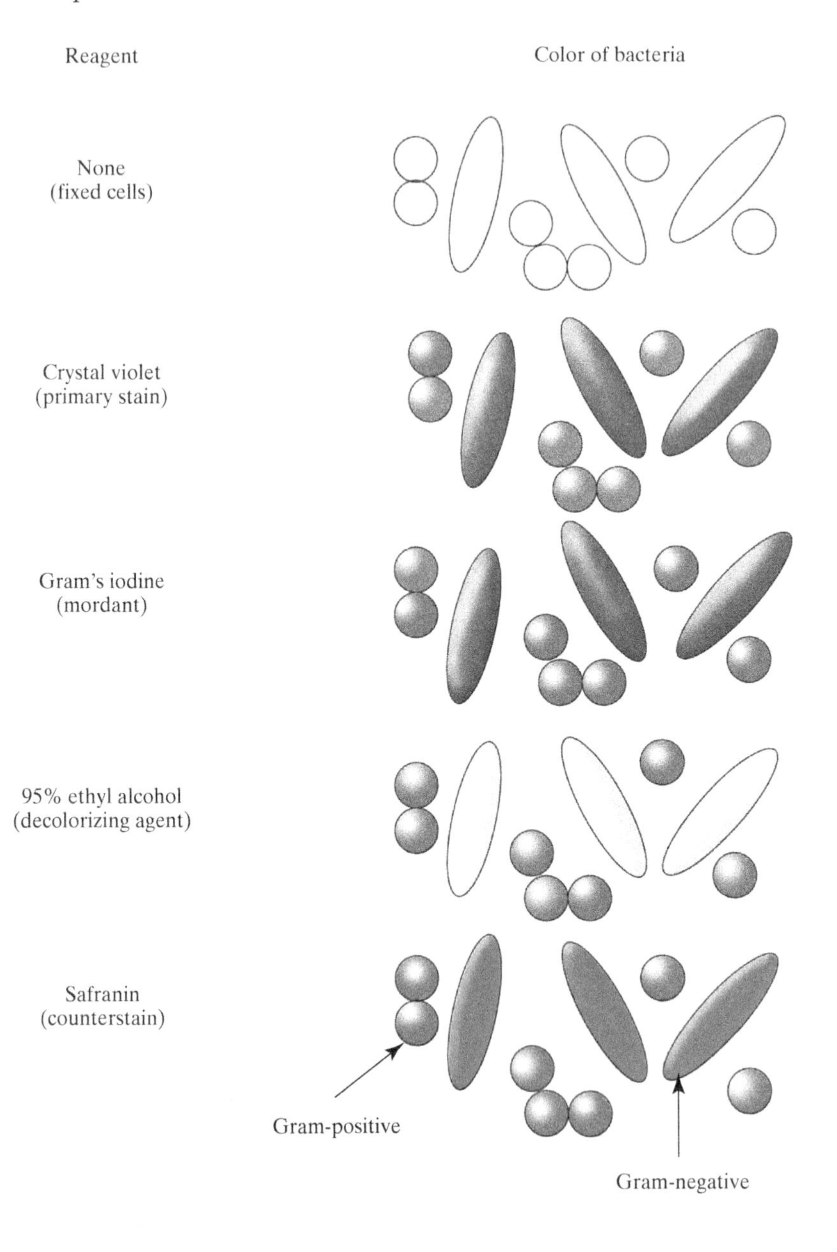

Staining Bacteria
Preparing an Acid-fast Stain

Purpose:

- To distinguish two groups of bacteria based on the lipid content in their cell walls: acid-fast and non–acid-fast.

Materials:

- microscope slide with fixed smear of bacteria
- stains in bottles with droppers:
  - carbolfuchsin (Ziehl-Neelsen *or* Kinyoun formulation)
  - acid-alcohol
  - methylene blue
- (Ziehl-Neelsen method) ring stand on which rests a 100 mL beaker with deionized water and boiling chips or beads heated to steaming over a burner
- (Ziehl-Neelsen method) pieces of paper toweling or filter paper, 1 × 2 in.
- staining rack over a tray or sink
- wash bottle with distilled or deionized water
  or
  rubber tubing connected to a faucet
- bibulous paper
- clothespin or forceps

Procedure:

Ziehl-Neelsen (hot method)

1. Place the slide on a beaker with boiling water.

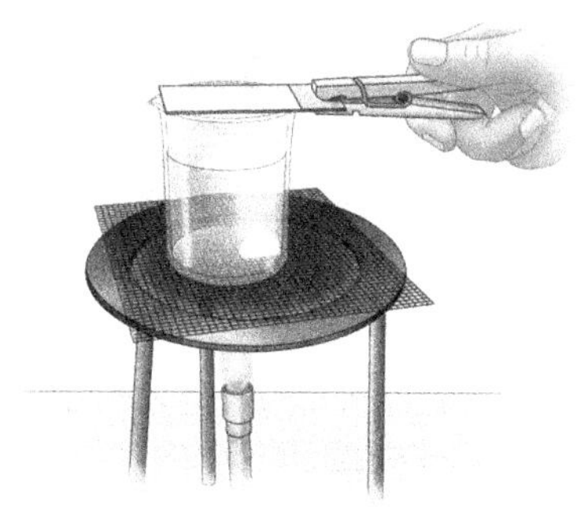

Kinyoun (cold method)

1a. Cover smear directly with Kinyoun's carbolfuchsin for 5 minutes. Skip to Step 5.

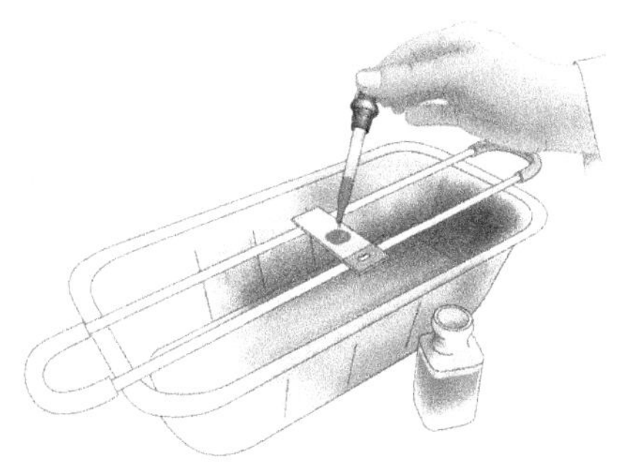

2. Place paper on the slide and saturate the paper with Ziehl-Neelsen carbolfuchsin.

3. Stain for 3–5 minutes. Add additional stain as it evaporates from the heat; do not allow the stain to dry.

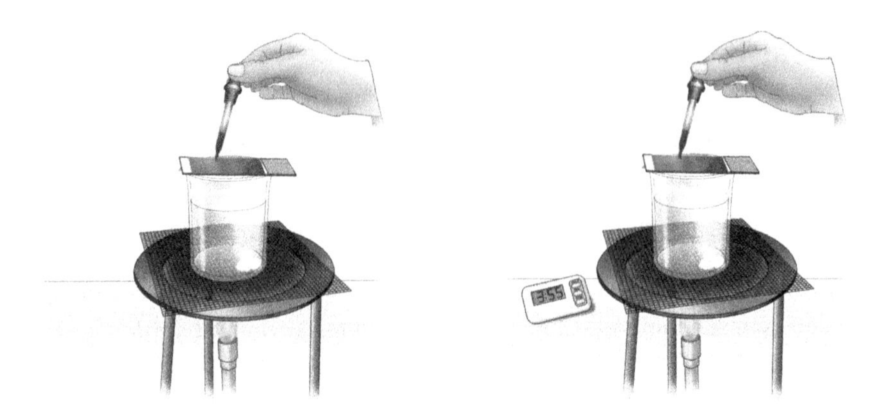

4. Use a forceps to remove the slide from the heat. Remove the paper and place it in a biohazard bag. Allow the slide to cool.

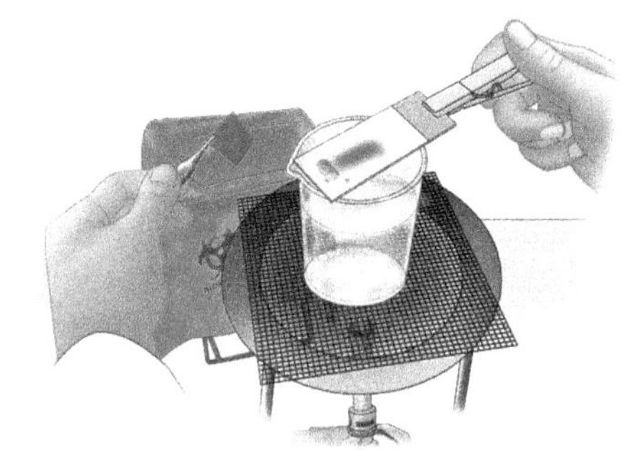

Continuing for both Ziehl-Neelsen and Kinyoun stains

5. Rinse the slide with water to remove excess stain.

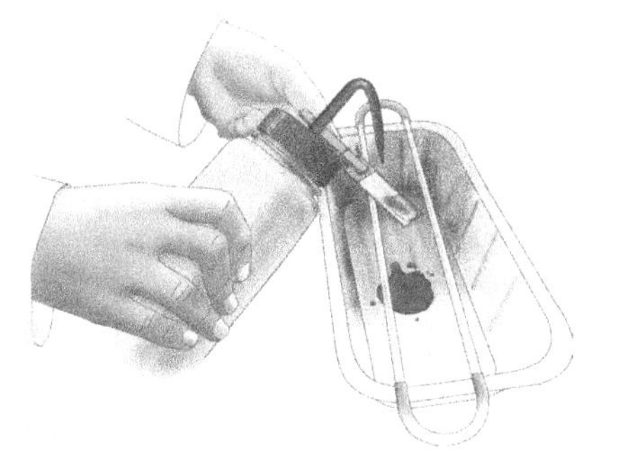

6. Decolorize with acid-alcohol. Hold slide at a 45° angle. Add decolorizing agent drop by drop until color stops running.

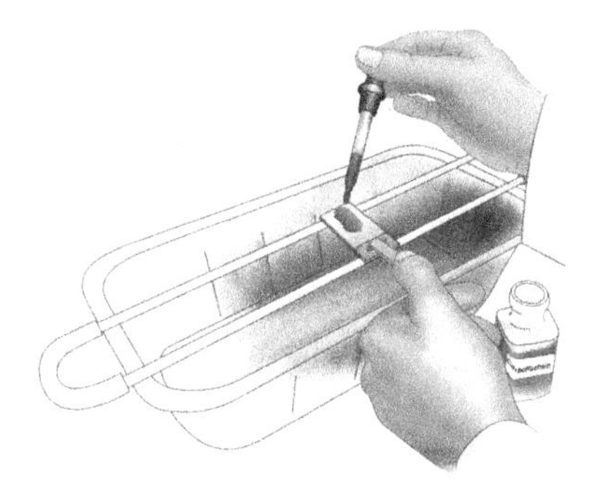

7. Immediately rinse the slide to remove the decolorizing agent.

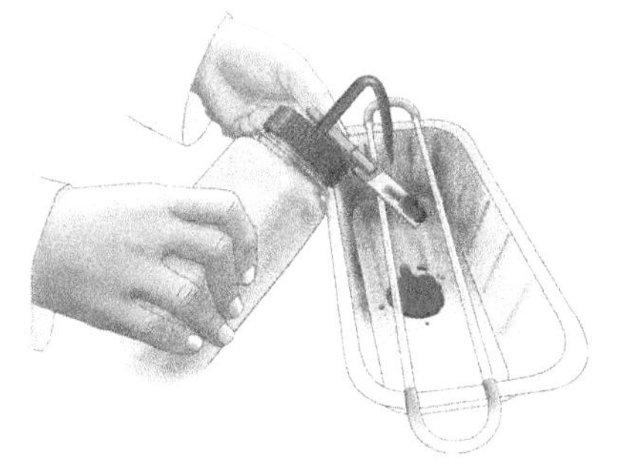

8. Cover the smear with methylene blue for 2 minutes.

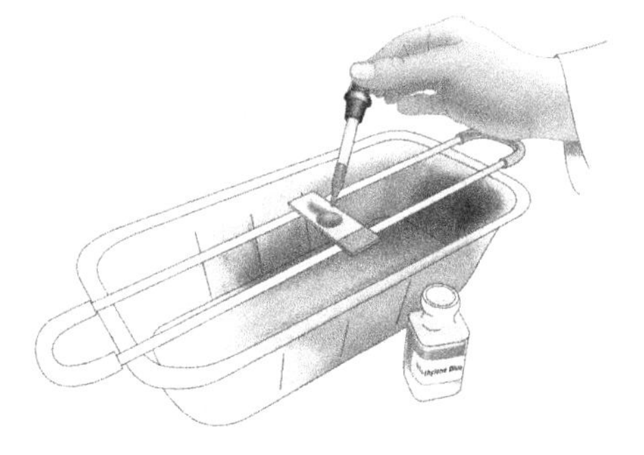

9. Rinse the slide with water to remove excess methylene blue.

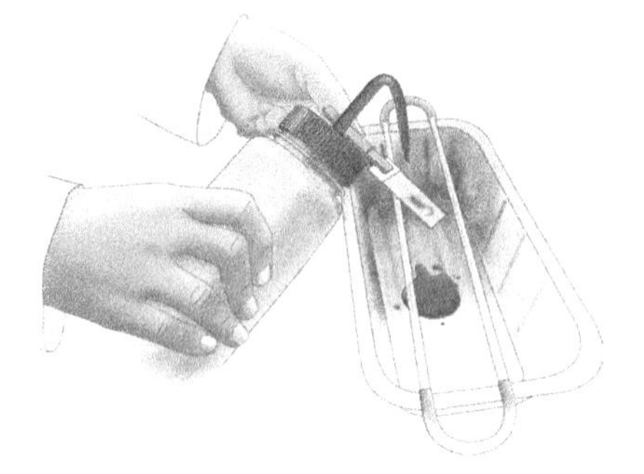

10. Blot water from the slide with pieces of bibulous paper.

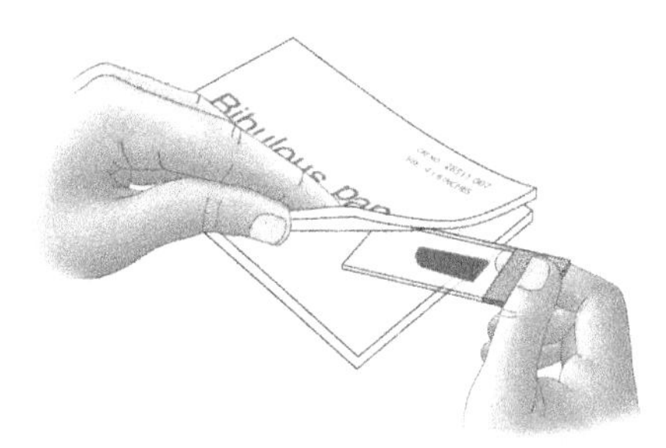

11. Examine the stained smear under the microscope using the oil immersion lens.

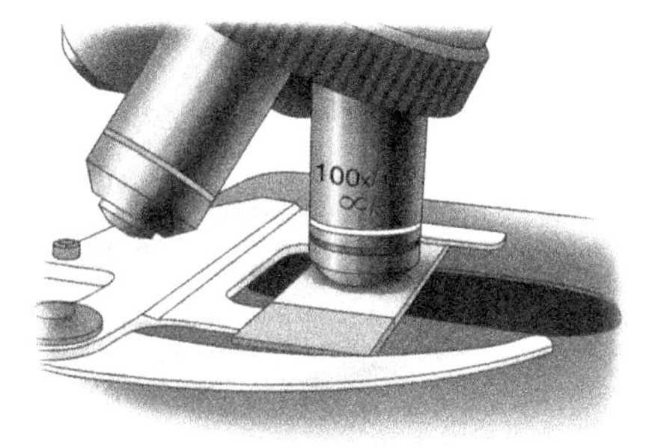

- *If you use boiling water, take extreme caution to avoid scalding your skin.*
- *If you use boiling water, wear safety glasses and use a clothespin or forceps to hold the slide.*
- *Keep the bottle and dropper of acid-alcohol away from an open flame—alcohol is flammable.*
- *Take care not to cut yourself if a glass slide breaks, especially when blotting the slide dry. Dispose of broken glass in the designated container.*
- *Be careful not to get stain on your clothing—it does not wash out. Clean stain from your fingers with stain-removing cream.*

Troubleshooting

- If the culture is less than 24 hours old, the cell walls of acid-fast bacteria may not have accumulated sufficient waxy material (mycolic acid) for a positive reaction.
- If too much decolorizing agent is applied, too much of the carbolfuchsin will wash out of the cell wall, giving a false non–acid-fast reaction.
- If you blot the slide too vigorously with bibulous paper, you may rub off the stained smear.

Appearance of cells in a fixed smear at each step in the acid-fast stain procedure

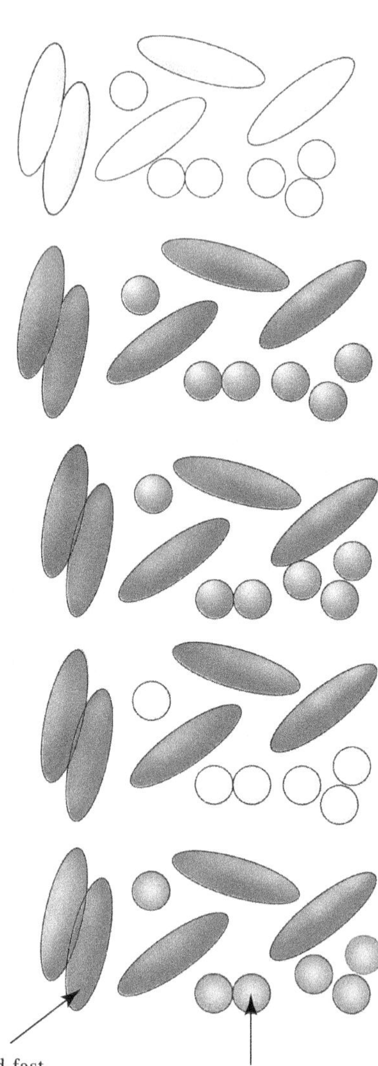

Staining Bacteria

Preparing a Negative Stain

Purpose:

- To view bacteria under the microscope by creating a dark or opaque background around transparent cells.
- To view bacterial cells that are difficult to stain with basic dyes.
- To view bacteria that are fragile easily distorted by heat-fixation.

Materials needed:

- clean microscope slide
- nigrosin stain
- staining rack over a tray or sink

Procedure:

1. Place a small drop of nigrosin near one end of a clean slide.

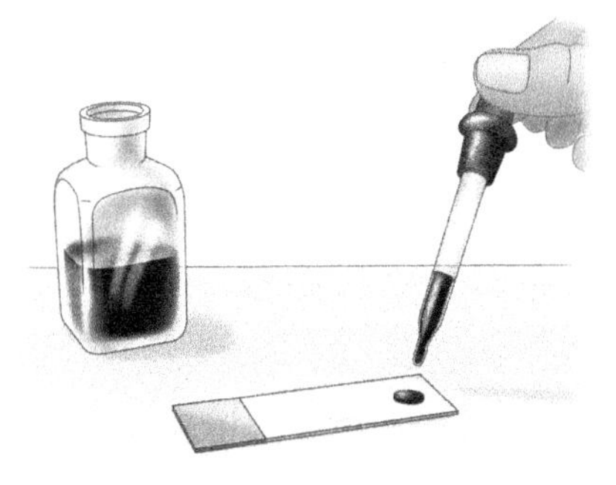

2. Use a sterilized inoculating loop to transfer a small amount of bacteria from an agar surface or a loopful of broth culture into the nigrosin drop. Mix well within that small diameter.

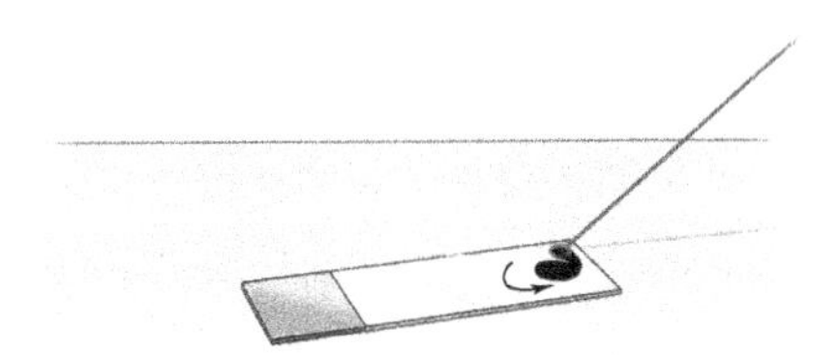

3. Flame the loop before putting it down.

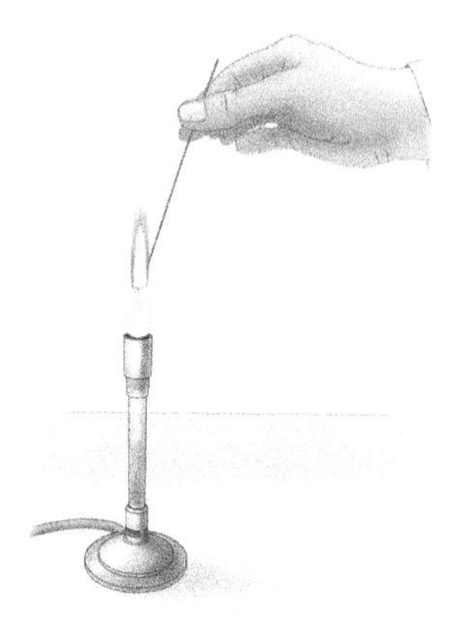

4. Touch the short edge of another clean microscope slide (spreader slide) at a 30–45° angle in the bacteria–nigrosin drop.

30–45°

5. After the bacteria–nigrosin drop has spread along the edge of the spreader slide, quickly push the slide to spread out the drop.

6. The resulting smear should include a thin film with a feathered edge at the trailing end.

7. Allow the smear to air dry completely.

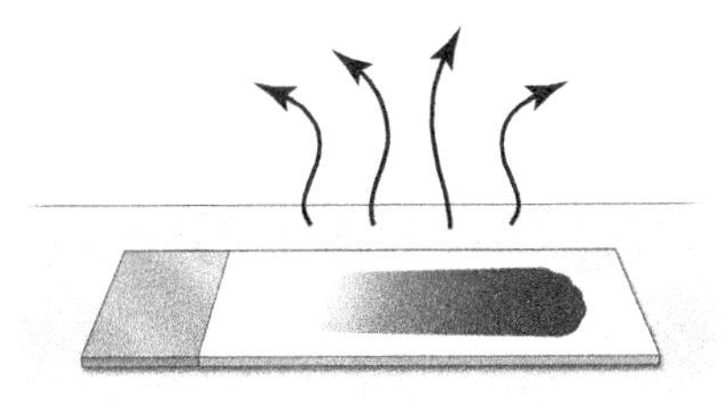

8. Examine the stained smear under the microscope using the oil immersion lens.

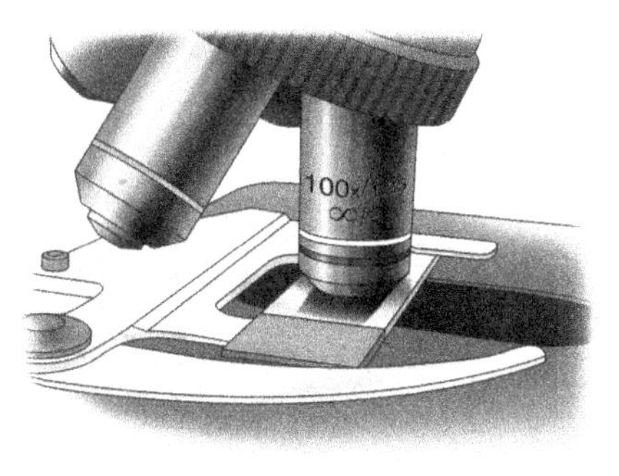

9. Dispose of the slide in a disinfectant solution.

- *The bacteria remain alive after the smear has been prepared because they were not killed by heat-fixation or by treatment with a dye. The slide must be placed in disinfectant or sterilized by autoclaving when you have finished with it.*
- *The oil immersion lens must be disinfected with 70% alcohol after you have finished examining the stain with the microscope. The lens paper should then be placed in a suitable biohazard container.*

Troubleshooting

- If the smear is too thick, it will have cracks and the cells will not be clearly outlined. The spreader slide may not have pulled the bacteria–nigrosin mixture evenly. You will have to prepare a new slide.
- If the bacteria–nigrosin mixture is pushed rather than pulled, the film will not be uniform.

STAINING BACTERIA

Preparing a Capsule Stain

Purpose:

- To view bacterial capsules or slime layers.

Materials needed:

- clean microscope slide
- nigrosin stain
- safranin or crystal violet
- staining rack over a tray or sink
- wash bottle with distilled or deionized water
  or
  rubber tubing connected to a faucet

Procedure:

1. Prepare a smear of bacteria in nigrosin as described in the procedure for a negative stain.

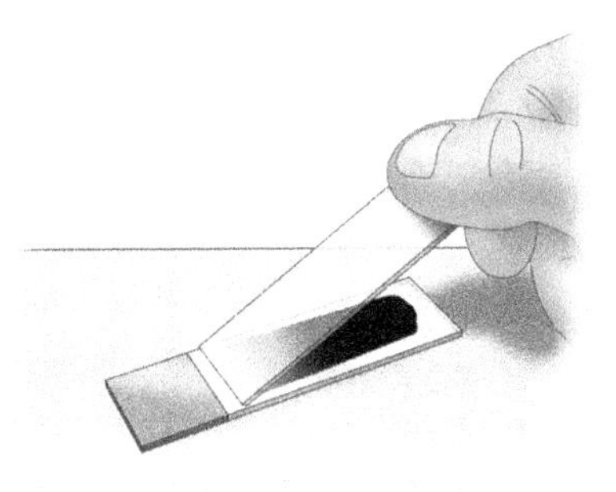

2. After allowing the spread smear to air dry, cover it with safranin or crystal violet.

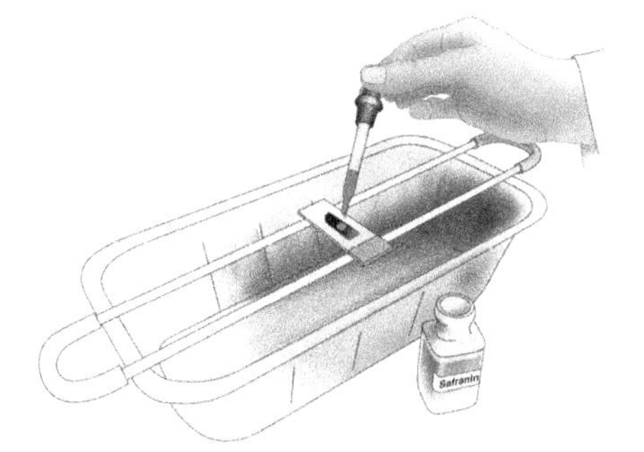

3. Gently wash off the excess stain. Avoid excess rinsing which will remove much of the smear.

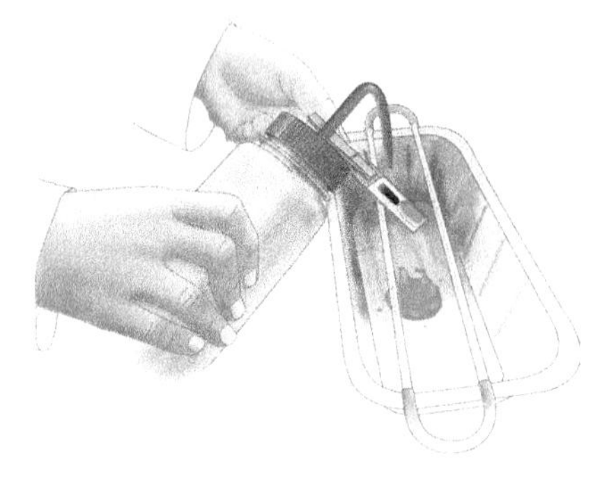

4. Blot water from the slide with pieces of bibulous paper.

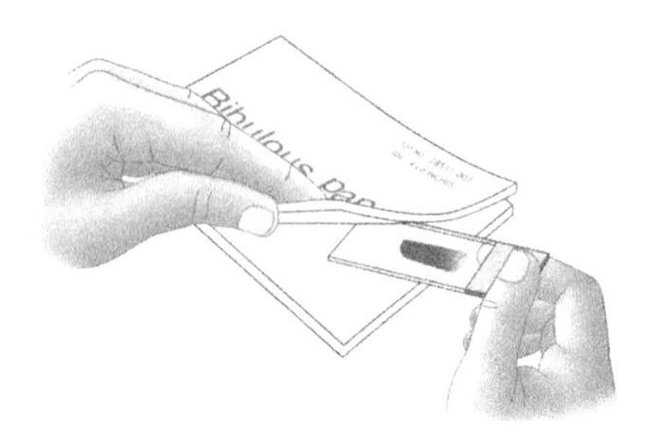

5. Examine the smear in the thin film under the microscope using the oil immersion lens.

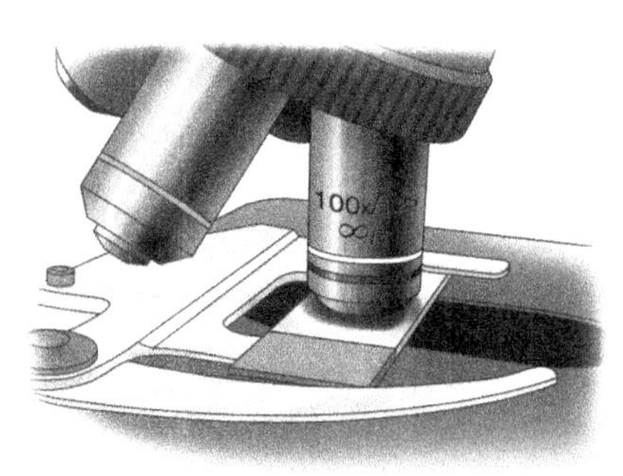

6. Dispose of the slide in a disinfectant solution.

- Position open flames so that you will not burn yourself. Tie back long hair back to prevent it catching on fire.
- Allow a flamed inoculating loop to cool before you remove bacteria from a culture. Spattering caused by a loop that is too hot can create an aerosol.
- Do not set a used inoculating loop down without sterilizing it—it will contaminate the work surface.
- Take care not to cut yourself if a glass slide breaks, especially when blotting the slide dry. Dispose of broken glass in the designated container.
- Be careful not to get stain on your clothing—it does not wash out. Clean stain from your fingers with stain-removing cream.

Troubleshooting

- If the smear is too thick, it will have cracks and the cells will not be clearly outlined. The spreader slide may not have pulled the bacteria–nigrosin mixture evenly. You will have to prepare a new slide.
- If the bacteria–nigrosin mixture is pushed rather than pulled, the film will not be uniform.

STAINING BACTERIA
Preparing an Endospore Stain

Purpose:

- To view bacterial endspores under the microscope.
- To observe the location of an endospore in a sporulating cell.

Endospores are formed by a few genera of gram-positive bacteria. Endospores are specialized for survival in unfavorable growth conditions, including heat and drying. Identifying endospore-forming bacteria is important in the microbiology of soil, food, and medicine. A thick protein coat that protects the endospore from toxic chemicals also prevents penetration of stains used to visualize bacteria. In a Gram stain or simple stain of a bacterial smear, a clear area within a stained cell denotes an endospore. However, heat will facilitate permeation of the stain malachite green into the spore coat. Malachite green will not wash out of the spore coat when the vegetative cell is stained with safranin.

Materials:

- slide with fixed smear of bacteria
- staining reagents in bottles with droppers:
  - malachite green
  - safranin
- staining rack over a tray or sink
- steaming deionized water in a 100 mL beaker with boiling chips or beads
  - hot plate for heating water

  or
  - ring stand with wire gauze and gas burner for heating water
- piece of filter paper, 1 × 2 in.
- wash bottle with distilled or deionized water

  or

  rubber tubing connected to a faucet
- bibulous paper
- clothespin or forceps

Procedure:

1. Place the slide on a beaker with boiling water. Staining must be done under a lab hood.

2. Place paper on the slide and saturate the paper with malachite green.

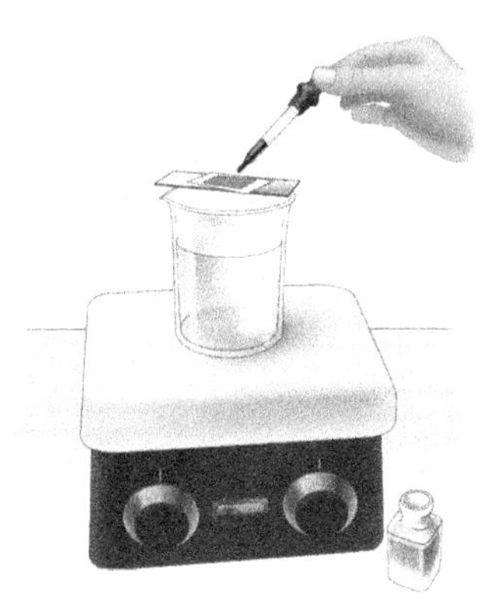

3. Stain for 5–6 minutes after the malachite green begins to steam. Add additional stain as it evaporates. Do not allow the stain to dry.

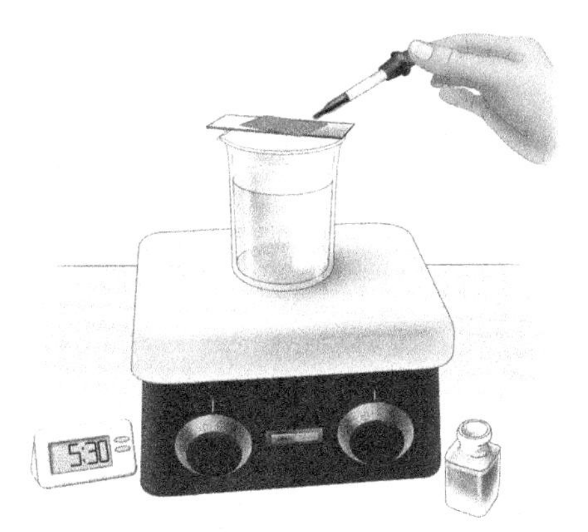

4. Use a forceps to remove the slide from the heat. Remove the paper and place it in a biohazard bag. Allow the slide to cool.

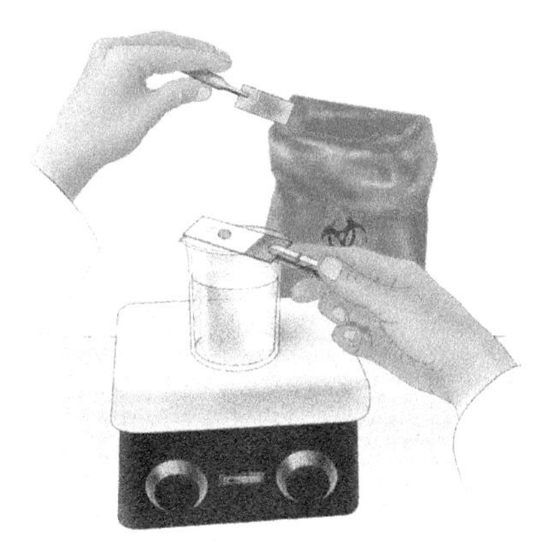

5. Rinse the slide with water for about 30 seconds to remove excess malachite green.

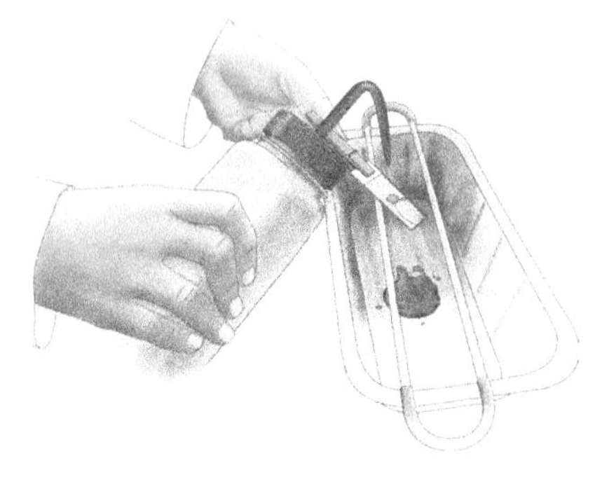

6. Cover the smear with safranin for 60–90 seconds.

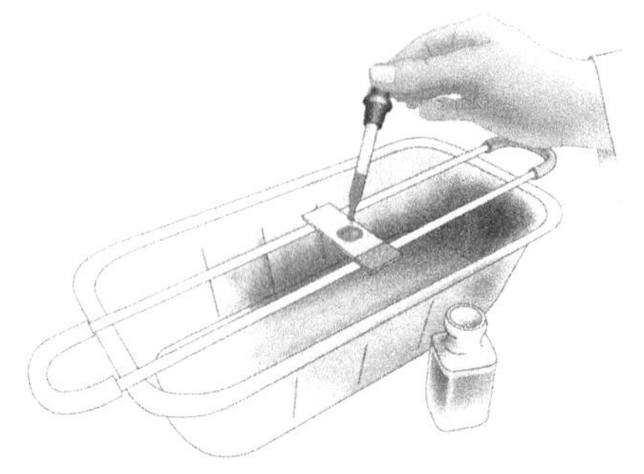

7. Rinse the slide with water to remove excess safranin.

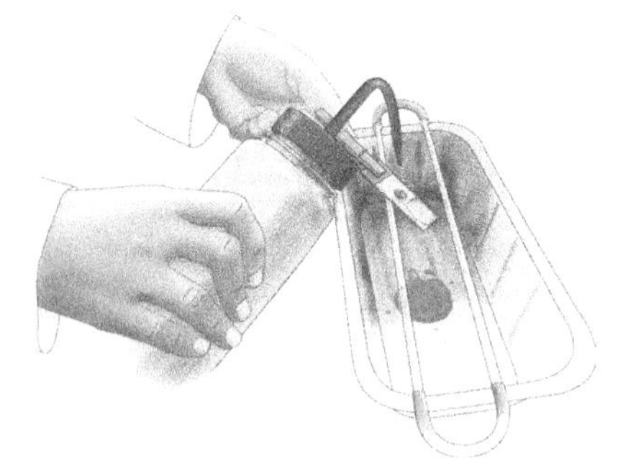

8. Blot water from the slide with pieces of bibulous paper.

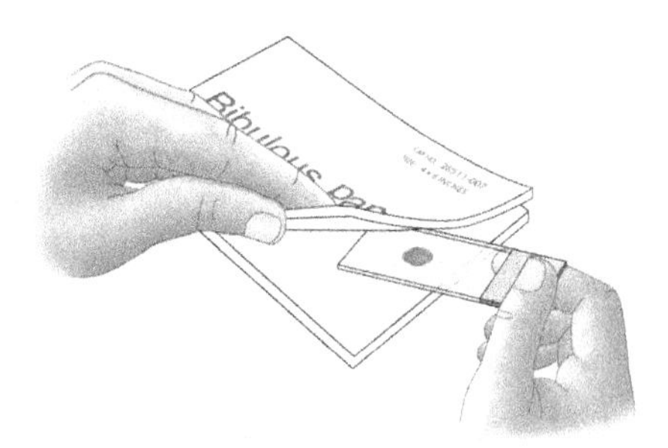

9. Examine the stained smear under the microscope using the oil immersion lens.

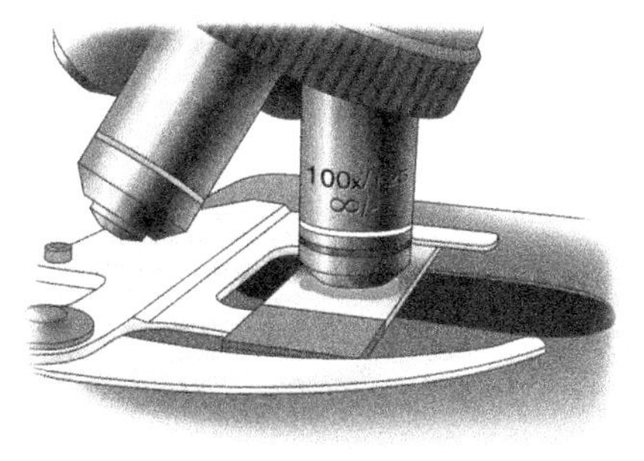

- *Always heat malachite green under a hood—the fumes are toxic.*
- *Wear eye protection as you heat the slide and malachite green. Use a clothespin or forceps to hold the hot slide.*
- *When you use boiling water, take extreme caution to avoid scalding your skin.*
- *Keep your lab coat buttoned to protect your clothes from malachite green—the stain does not wash out.*
- *Clean stain from your fingers with stain-removing cream*

Troubleshooting

- Use gentle steaming to heat the slide—do not allow the stain to boil.
- While steaming malachite green on the smear, add drops of malachite green to prevent drying of the stain.
- Thoroughly rinse the slide with water after staining with malachite green. Otherwise, the vegetative cells will not be stained with the red-colored counterstain safranin.
- If you blot the slide too vigorously with bibulous paper, you may rub off the stained smear.

Appearance of cells in a fixed smear at each step in the endospore stain procedure

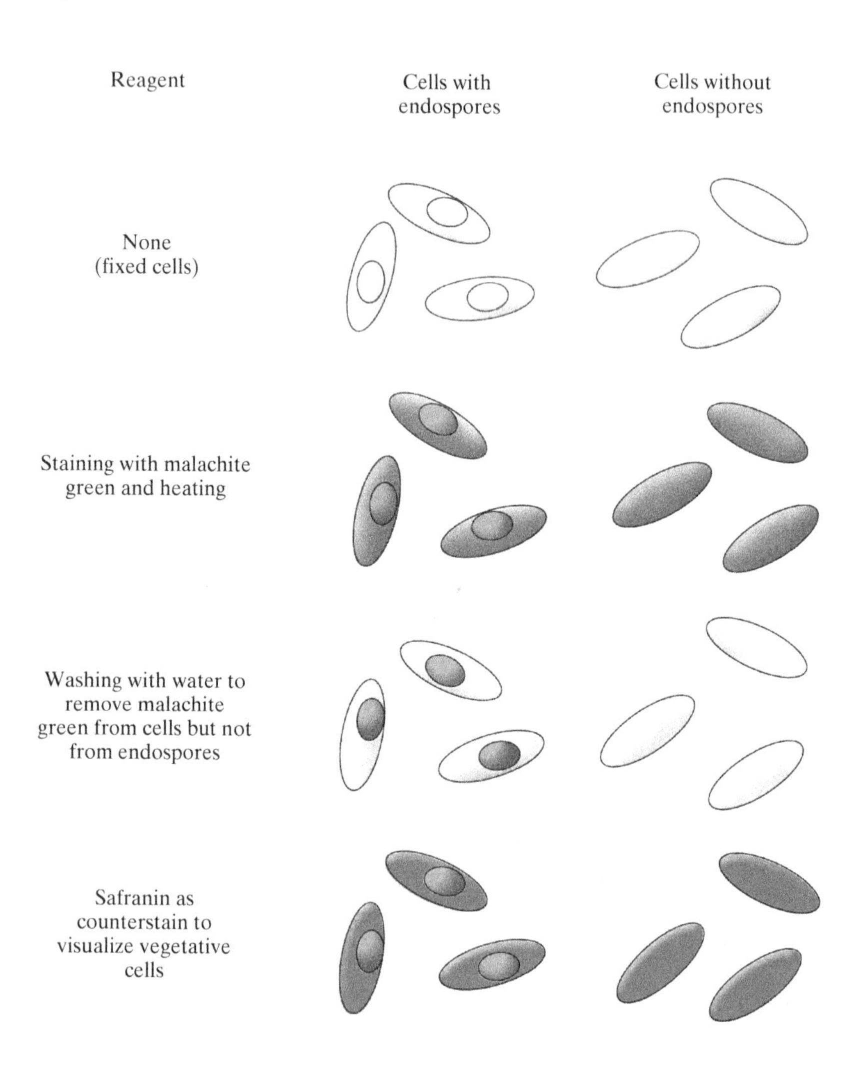

Measuring Microscopic Cells

The size of a microorganism is measured in micrometers, or µm. To measure such a minute size, a calibrated ruler on a glass disk called an **ocular micrometer** (my krah' meh ter) is placed in an ocular lens. The distance between each engraved line in the ocular micrometer spans one **ocular division (OD)**. To determine the exact distance measured by an OD, a **stage micrometer** is used for calibration. It has engraved lines that are exactly 10 µm apart. For each objective lens, a calibration factor for the distance (µm) measured by 1 OD is established. Then, with the calibration factor, the ocular micrometer is used as a ruler to measure the dimensions of the microorganisms.

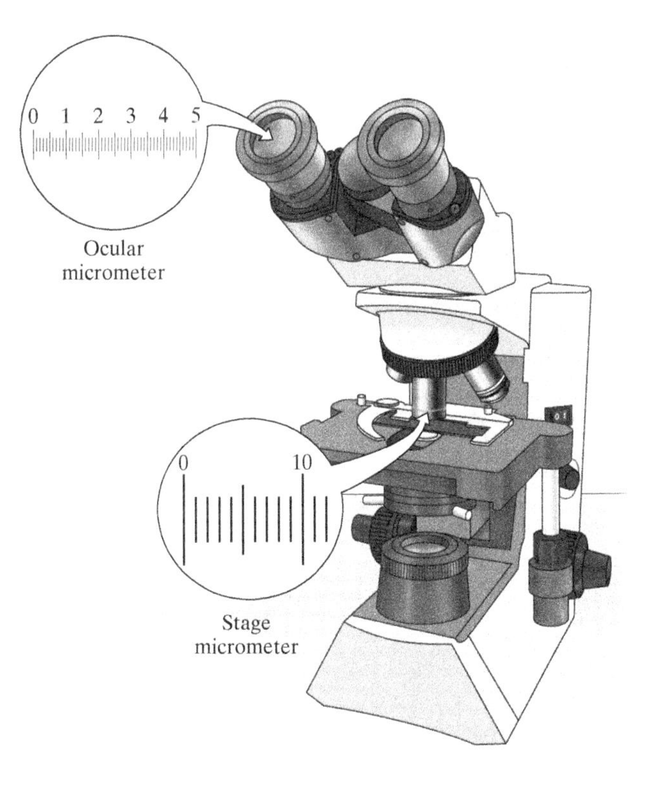

Materials:

- microscope with ocular micrometer in an ocular lens
- stage micrometer
- slides of stained microorganisms
- immersion oil

Procedure:

1. Center the stage micrometer on the stage over the light coming through the condenser.

2. With an objective lens clicked into place, use the coarse and then fine focus knobs to bring the engraved lines on the stage micrometer into sharp view. Adjust the light intensity if necessary.

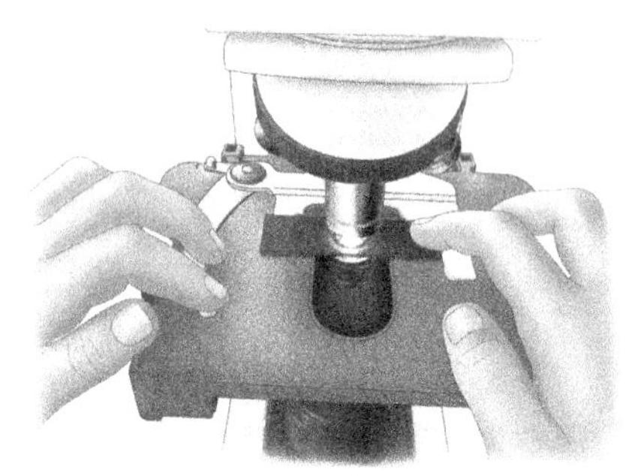

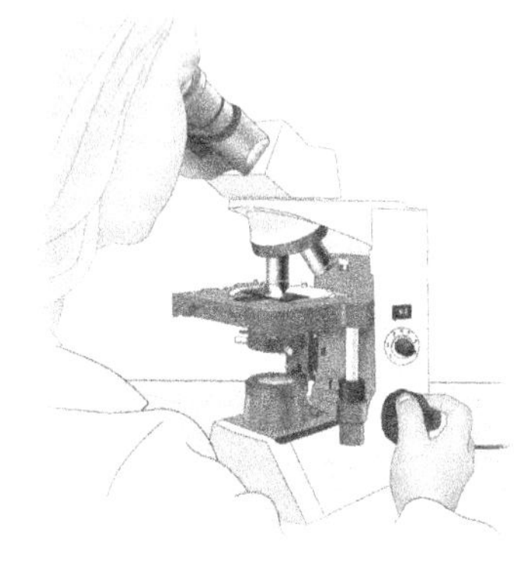

3. Rotate the ocular lens until the engraved lines on the ocular micrometer are parallel with those on the stage micrometer.

4. Use the mechanical stage adjustment knobs to move the stage micrometer until its left-end line is superimposed on the left-end line on the ocular micrometer.

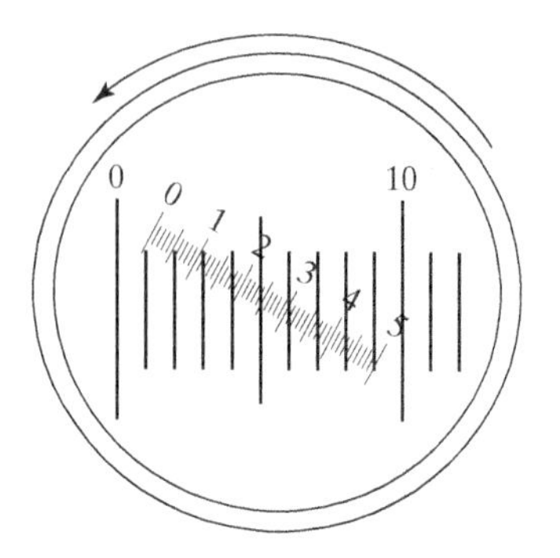

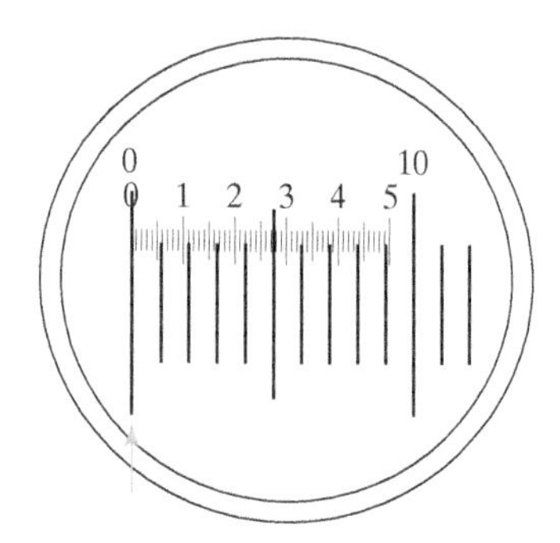

5. Look to the right for a line on the ocular micrometer that is exactly superimposed over a line on the stage micrometer.

6. Count the number of ocular divisions (OD) that span the actual distance measured on the stage micrometer.

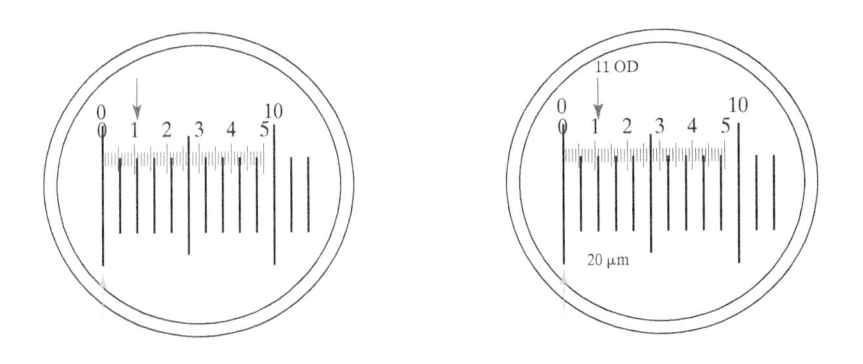

7. Calculate the calibration factor for the objective lens.
In Step 6, 11 OD measures 20 μm on the stage micrometer.

$$1 \text{ OD} = \frac{20\ \mu\text{m}}{11 \text{ OD}}$$

$$1 \text{ OD} = 1.8\ \mu\text{m}$$

8. Calibrate the ocular micrometer for each of the objective lenses. Remember to apply immersion oil for the 100X lens.

Lens	**Calibration factor**
10X	_____ OD/μm
40X	_____ OD/μm
100X	_____ OD/μm

9. Use lint-free lens paper to wipe the oil from the stage micrometer and the lens.

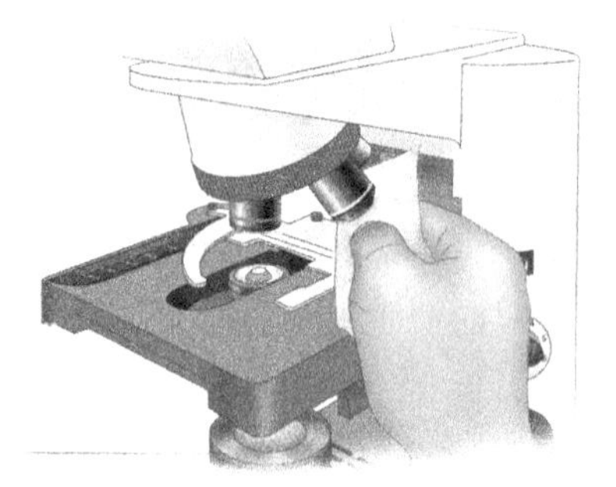

10. To measure the size of a bacterium on a slide, count the number of ODs that span the length or width of a cell. Then multiply this number by the calibration factor for the objective lens.

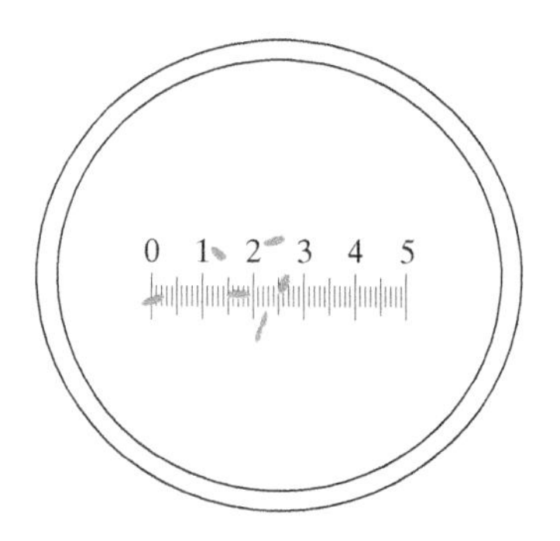

Size = # of ODs X Calibration Factor

In this figure, the horizontal cell covers 6 ODs (14 to 20). Therefore:

Length = 6 X 1.8 = 10.8

Hydrolytic (Digestive) Enzymes

Starch Hydrolysis Test

Purpose:

- To distinguish among bacteria in terms of their ability to hydrolyze (digest) starch.

Plants store glucose as very large molecules called starch. Starch is made up of amylose, a long unbranched polymer of several hundred glucose subunits, and amylopectin, a branched polymer. When plants die, bacteria in the soil digest the starch in the plant cells. The bacteria use an extracellular enzyme called **amylase** to hydrolyze the bonds that link the glucose subunits. The products glucose and maltose are transported across the bacterial plasma membrane, where they are used for energy and construction of other biomolecules.

$$\text{starch} \xrightarrow{\text{amylase}} \text{glucose} + \text{maltose}$$

(iodine) (purple) (no color)

Starch in a medium is detected by adding an iodide/iodine solution. When iodine reacts with starch, a purple-black color develops. If bacteria growing on starch agar produce amylase, all of the starch around the growth will be consumed after incubation. When iodine is added, no color change around the bacterial growth means that starch is no longer present. This is a positive test for starch hydrolysis.

Materials:

- starch agar plate
- Gram's iodine (also used for Gram stain)
- pure culture of bacteria in a broth culture or on an agar slant

Procedure:

1. Label the bottom of a starch agar plate with the name of the bacterium, your name or initials, and the date.
2. Use aseptic technique to inoculate the starch agar plate by streaking a short line on the agar surface with a loopful of broth culture or a small inoculum from an agar slant culture.

3. Incubate the inoculated plate upside-down at room temperature for several days, or at 35° C for 48 hours.
4. After incubation, flood the starch agar surface with Gram's iodine. Wait 30 seconds to 1 minute for a purple-black color to develop that indicates where starch is present in the agar.
5. A clear area around a bacteria growth is a positive test for starch hydrolysis. If color appears around the growth, the test is negative.
6. Discard the plate in the designated place for sterilization. Be careful that you do not spill the iodine solution, as it contains viable bacteria.

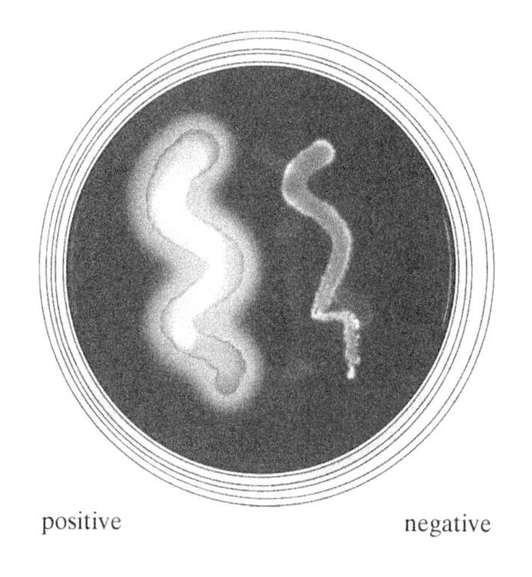

Interpretation of a starch hydrolysis test

Troubleshooting

- If the Gram's iodine appears yellow rather than a brownish-orange, it has aged and must be replaced with fresh reagent.
- Check the plate for distribution of color immediately—the color fades quickly.

HYDROLYTIC (DIGESTIVE) ENZYMES

Casein Hydrolysis Test

Purpose:

- To distinguish among bacteria in terms of their ability to hydrolyze (digest) casein, the major protein in milk.

Many bacteria secrete enzymes to hydrolyze proteins in their immediate surroundings. The resulting amino acids and peptides (short chains of amino acids) are then transported across cell membranes where they are used to construct bacterial proteins, to serve as carbon and nitrogen sources, and to serve as an energy source.

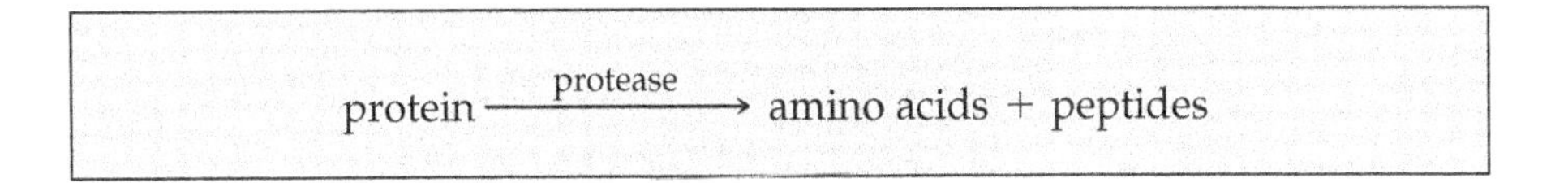

$$\text{protein} \xrightarrow{\text{protease}} \text{amino acids} + \text{peptides}$$

Casein, the predominant protein in milk, can be used as a substrate to assess the production of proteinases (or proteases) by certain bacteria. When mixed with agar, casein forms a white colloid with calcium ions. When bacteria secrete **caseinase**, a specific protease, casein molecules are digested and the area around the bacterial growth becomes clear. This clear zone is a positive test. If the bacteria do not secrete caseinase, the medium remains white around the bacterial growth.

Materials:

- skim milk agar plate
- pure culture of bacteria in a broth culture or on an agar slant

Procedure:

1. Label the bottom of the skim milk agar plate with the name of the bacterium, your name or initials, and the date.
2. Use aseptic technique to inoculate the skim milk agar plate by streaking a short line on the agar surface with a loopful of broth culture or a small inoculum from an agar slant culture.
3. Incubate the inoculated plate upside-down at 35° C for 48 hours.

4. After incubation, examine the plate for a clear zone around the bacterial growth on the plate. This is a positive test. If the area around the bacteria remains white, the test is negative.

5. Discard the plate in the designated place for sterilization.

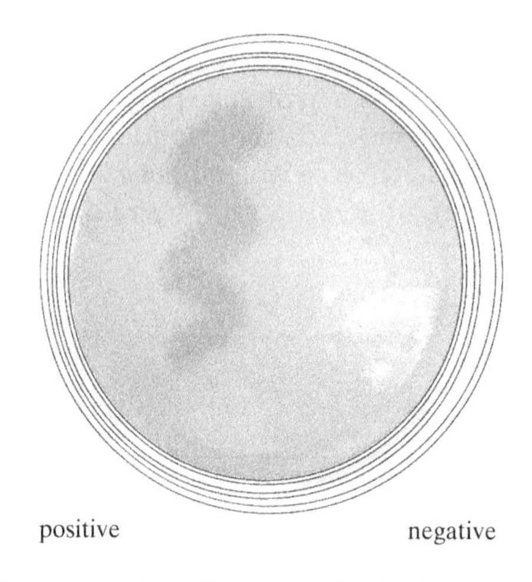

Interpretation of a casein hydrolysis test

HYDROLYTIC (DIGESTIVE) ENZYMES

Gelatin Hydrolysis Test

Purpose:

- To determine the ability of bacteria to hydrolyze (digest) gelatin.

Bacteria secrete protein-digesting enzymes that degrade foreign proteins around them. One animal protein that is a substrate for some bacterial proteinases is **gelatin**. Gelatin is a processed form of collagen, an important protein in connective tissues. It is produced by extraction from the skin and bones of hogs and cows. When heated in water and then cooled (< 20° C), gelatin forms into a semi-solid gel.

Some bacteria can hydrolyze gelatin with the extracellular enzyme **gelatinase**. As a protease, gelatinase cleaves multiple peptide bonds in gelatin, causing the gel to liquefy. If a nutrient gelatin tube inoculated with bacteria remains liquid after cooling to 4° C (melting ice or the refrigerator), the bacteria are positive for gelatinase.

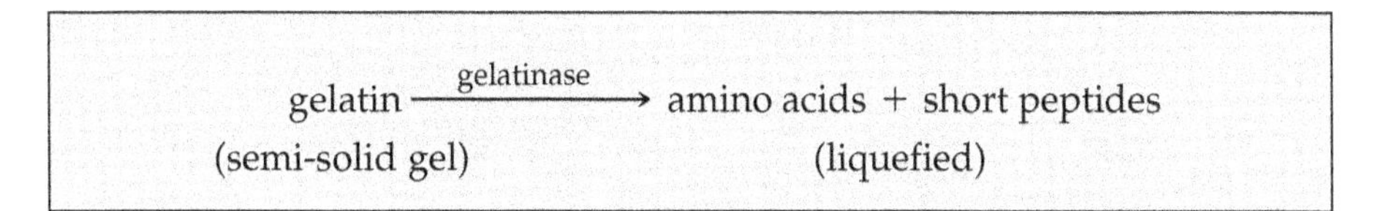

Materials:

- nutrient gelatin deep tube
- pure culture of bacteria in a broth culture or on an agar slant
- refrigerator or an ice bucket with melting ice

Procedure:

1. Label the nutrient gelatin deep tube with the name of the bacterium, your name or initials, and the date.

2. Use aseptic technique to inoculate the nutrient gelatin deep tube by stabbing with an inoculating needle carrying bacteria from a broth culture or an agar slant culture.

3. Incubate the inoculated tube at either room temperature or at 35° C. Incubate an uninoculated tube as a control.

4. After 48 hours, place the tubes in a melting ice bath or in the refrigerator for 15 minutes. Check each for liquifaction by tilting the tube slightly. If the medium in the inoculated tube remains liquid, the test for gelatinase is positive if the control tube has solidified. Because some bacteria hydrolyze gelatin at a slow rate, incubate the tubes for several more days if the inoculated medium is still a gel. Repeat cooling before you look for liquefying of the medium.

5. Discard the tubes in the designated place for sterilization.

positive

negative

Interpretation of a gelatin hydrolysis test

- *Take care not to tilt the tube too much so that the inoculated, liquefied medium runs out of the tube and contaminates the environment.*

HYDROLYTIC (DIGESTIVE) ENZYMES
Fat (Triglyceride) Hydrolysis Test

Purpose:

- To determine the ability of a bacterium to hydrolyze (digest) a triglyceride, a type of lipid.

Some bacteria live in habitats where animal or vegetable lipids are present, such as human skin or soil with dead organisms. Most of these lipids consist of **triglycerides**, water-insoluble molecules made from three fatty acids linked to glycerol. To use triglycerides for food and energy, bacteria must secrete hydrolytic enzymes called **lipases**. Hydrolysis of triglycerides yields fatty acids and glycerol, which the bacteria take up and use for energy and for constructing their own biomolecules.

$$\text{triglyceride} \xrightarrow{\text{lipase}} \text{fatty acids} + \text{glycerol}$$

The test medium for triglyceride hydrolysis is tributyrin agar. Tributyrin is an oily triglyceride that is emulsified with melted agar. The cooled medium appears cloudy. When bacteria growing on the medium secrete lipases that hydrolyze tributyrin, a clear zone appears around the growth. This represents a positive test for lipid hydrolysis. If no clearing is seen, the test is negative.

Materials:

- tributyrin agar plate
- pure culture of bacteria in a broth culture or on an agar slant

Procedure:

1. Label the bottom of a tributyrin agar plate with the name of the bacterium, your name or initials, and the date.

2. Use aseptic technique to inoculate the tributyrin agar plate by streaking a short line on the agar surface with a loopful of broth culture or a small inoculum from an agar slant culture.

3. Incubate the inoculated plate upside-down at 35° C for 48 hours.

4. After incubation, examine the plate for a clear area around the bacteria growth. This is a positive test for triglyceride hydrolysis. If no clearing is seen around the growth, the test is negative.

5. Discard the plate in the designated place for sterilization.

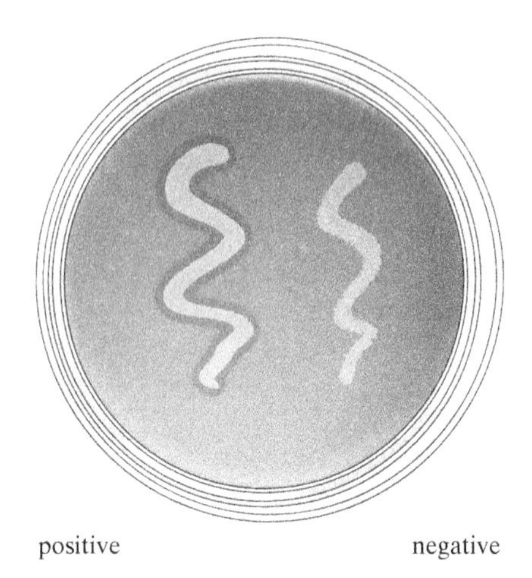

Interpretation of a fat (triglyceride) hydrolysis test

HYDROLYTIC (DIGESTIVE) ENZYMES

DNA Hydrolysis Test

Purpose:

- To detect the ability of a bacterium to hydrolyze (digest) DNA.

Some bacteria secrete the enzyme **DNase** that hydrolyzes many of the linkages between the nucleotides in DNA, leaving small DNA fragments. DNase-producing bacteria may use the enzyme to spread through infected tissues "clogged" with viscous DNA released from dead host cells.

To detect DNA hydrolysis by bacteria growing on a DNA agar plate, 1N hydrochloric acid (HCl) is added to the plate. The weak acid causes the intact DNA in the agar to appear cloudy. If the bacteria secreted DNase, no large DNA molecules remain around the growth. A clear zone around the bacterial growth after addition of HCl is a positive test for DNA hydrolysis.

A similar DNA agar includes the dye methyl green. A positive test for this medium is a zone with no color around the bacterial growth. No HCl is added.

Materials:

- DNA agar plate
- 1N hydrochloric acid (HCl) in a bottle with a dropper or Pasteur pipette and rubber bulb.
- pure culture of bacteria in a broth culture or on an agar slant

Procedure:

1. Label the bottom of a DNA agar plate with the name of the bacterium, your name or initials, and the date.
2. Use aseptic technique to inoculate the DNA agar plate by streaking a short line on the agar surface with a loopful of broth culture or a small inoculum from an agar slant culture.
3. Incubate the inoculated plate upside-down at 35° C for 24 hours.
4. After incubation, flood the plate with 1N HCl. Examine the plate for a clear area around the bacteria growth. This is a positive test for DNase. If no clearing is seen around the growth, the test is negative.

5. Discard the plate in the designated place for sterilization.

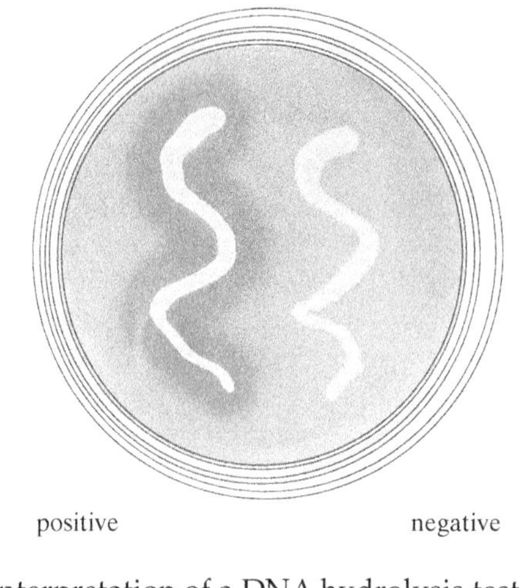

Interpretation of a DNA hydrolysis test

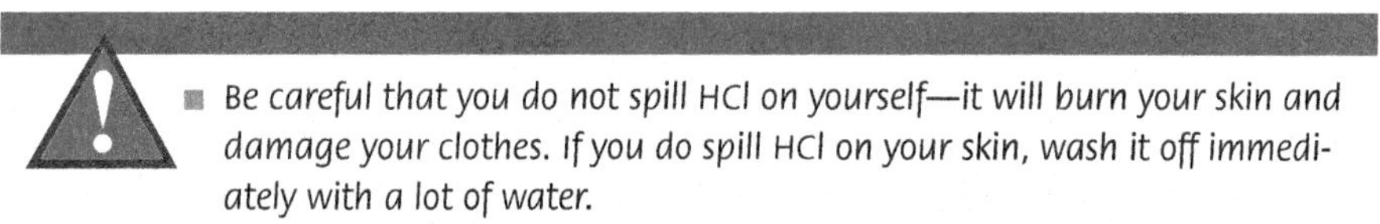

- *Be careful that you do not spill HCl on yourself—it will burn your skin and damage your clothes. If you do spill HCl on your skin, wash it off immediately with a lot of water.*
- *When you dispose of the plate, take care that you do not spill HCl.*

UTILIZATION OF CARBOHYDRATES

Fermentation of Carbohydrates (Durham Tube)

Purpose:

- To determine the ability of some bacteria to ferment a particular carbohydrate.

Heterotrophic bacteria often use sugars in fermentation pathways to obtain biologically usable energy. Organic acids, alcohols, and gases accumulate as waste products: waste molecules vary depending on the bacteria. In some fermentation pathways, only organic acids and/or alcohols are byproducts. In others, CO_2 and H_2 are released as gases in addition to the organic acids and alcohols.

In this test, acid production is identified by a change in the color of the phenol red, a pH indicator that is included in the medium. In acid pH, phenol red is yellow. To collect gases, an inverted smaller tube (a Durham tube) is placed in the medium.

Utilization of disaccharides, such as lactose and sucrose, by bacteria requires the production of a specific hydrolytic enzyme. Once the disaccharide is digested, the resulting monosaccharides are fermented. For example, a bacterium identified as a "lactose-fermenter" must synthesize the enzyme **lactase,** also called β-galactosidase.

$$\text{lactose} \xrightarrow{\text{lactase}} \text{galactose} + \text{glucose}$$
$$\downarrow \text{fermentation}$$
$$\text{organic acids} + (\text{alcohols}) + (CO_2/H_2)$$

The set of disaccharide-digesting enzymes made by a particular bacterium contributes to the microbe's unique molecular "fingerprint."

Materials:

- tube of phenol red–sugar broth with Durham tube: sugars can include glucose, lactose, sucrose or maltose; also possible are the sugar alcohols mannitol and sorbitol
- pure culture of bacteria in a broth culture or on an agar slant

Procedure:

1. Label the tube containing the phenol red–sugar broth and Durham tube with the name of bacterium you are testing, your name or initials, and the date.
2. Use aseptic technique to inoculate the phenol red–sugar broth tube with either a loopful of bacteria from a broth culture or a small inoculum from an agar slant culture.
3. Incubate the inoculated tube at 35° C for no more than 24 hours.
4. After incubation, examine the tube for acid production or acid and gas production. A yellow color is a positive test for acid production. An orange or red color is negative. A gas bubble trapped in the Durham tube is a positive text for gas production.
5. Discard the tube in the designated place for sterilization.

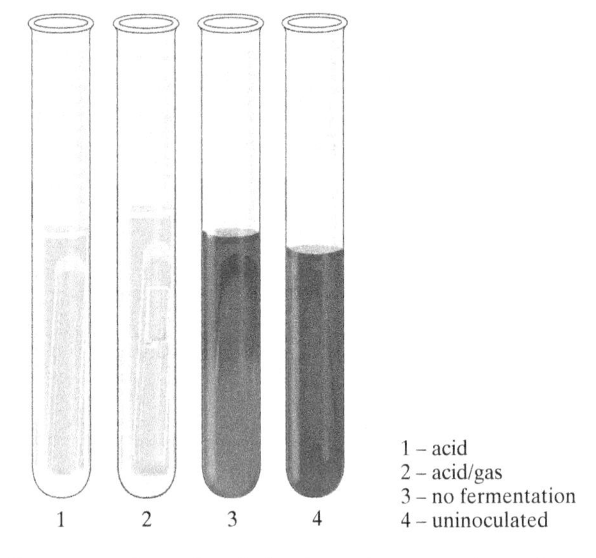

Interpretation of carbohydrate fermentation tubes

Troubleshooting

- If the tubes are examined after 24 hours, you may see a false negative. After fermenting the sugar, some bacteria will degrade amino acids in proteins included in the medium and release ammonia. This will raise the pH and change the color of the medium back to red.

Utilization of Carbohydrates
Methyl Red Test (Mixed Fermentation)

Purpose:

- To determine the ability of some bacteria to ferment glucose via mixed-acid fermentation.

Products of **mixed-acid fermentation** of glucose include significant amounts of organic acids.

$$\text{glucose} \xrightarrow{\text{mixed-acid fermentation}} \text{acetic acid} + \text{succinic acid} + \text{formic acid} + \text{gas}$$

These acids lower the pH of the medium to < 5. When the pH indicator **methyl red** is added to the medium and the pH is 4.5 and below, the methyl red reagent remains red in color. This is a positive test. At higher pH values (less acid present), the color may be orange or yellow. These colors denote a negative test.

Materials:

- methyl red/Voges-Proskauer (MR-VP) broth tube
- pure culture of bacteria in a broth tube or on an agar slant
- methyl red (pH indicator) in a dropper bottle or in a bottle with a Pasteur pipette and bulb

Procedure:

1. Label the tube of MR-VP broth with the name of the bacterium you are testing, your name or initials, and the date.
2. Use aseptic technique to inoculate the MR-VP tube with either a loopful of bacteria from a broth culture or a small inoculum from an agar slant culture.
3. Incubate the inoculated tube at 35° C for at least 48 hours.
4. After incubation, add 5–6 drops of the methyl red to the tube. Gently swirl the tube to mix the broth culture and the pH indicator.

5. Read the reaction immediately. A red color is a positive test for the mixed-acid fermentation pathway. An orange or yellow color is a negative test.

6. Discard the tube in the designated place for sterilization.

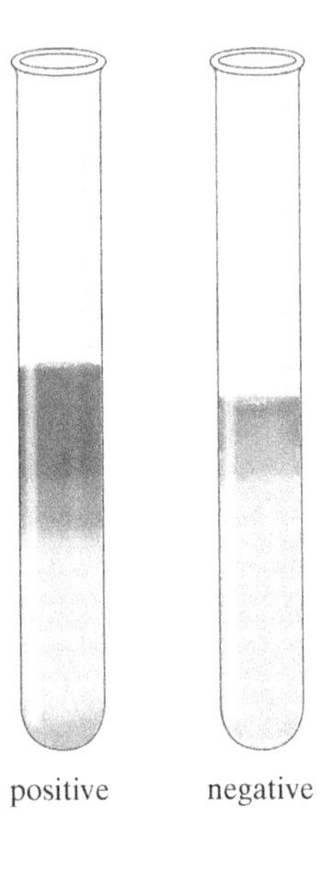

Interpretation of a methyl red (mixed fermentation) test

Troubleshooting

- If the tube is not gently swirled after you add the drops of methyl red, you may see a false positive. The pH indicator needs to be mixed to be able to react with available H^+ in the culture.
- If the tube is incubated for less than 48 hours, an insufficient level of acidic end products may accumulate in the broth. The result would be a false negative.
- Do not use the same tube in which you did a Voges-Proskauer test to do a methyl red test. The reagents for each test will interfere with each other.

UTILIZATION OF CARBOHYDRATES

Voges-Proskauer Test (Butanediol Fermentation)

Purpose:

- To determine the ability of some bacteria to ferment glucose via butanediol fermentation.

Some bacteria ferment glucose via the **butanediol fermentation** pathway. The precursor for the neutral alcohol 2,3 butanediol is acetylmethylcarbinol, also known as **acetoin**.

butanediol fermentation
glucose $\rightarrow\ \rightarrow$ α-acetolactate $\rightarrow\ \rightarrow$ acetoin $\longrightarrow$ 2,3 butanediol

While organic acids are also products of this pathway, they are not present in significant amounts after 48 hours of incubation.

Voges-Proskauer (VP) reagents react with acetoin in the presence of O_2 to form a red product. This red color denotes a positive VP test. No color change after addition of the VP reagents denotes a negative test.

Materials:

- methyl red/Voges-Proskauer (MR-VP) broth tube
- pure culture of bacteria in a broth tube or on an agar slant
- Barritt's (VP) reagent A (1% α-naphthol in ethanol) in a dropper bottle or in a bottle with a Pasteur pipette and a bulb
- Barritt's (VP) reagent B (40% KOH) in a dropper bottle or in a bottle with a Pasteur pipette and a bulb
- disposable gloves

Procedure:

1. Label the MR-VP broth tube with the name of the bacterium you are testing, your name or initials, and the date.
2. Use aseptic technique to inoculate the MR-VP tube with either a loopful of bacteria from a broth culture or a small inoculum from an agar slant culture.
3. Incubate the inoculated tube at 35° C for at least 48 hours.
4. After incubation, add 15 drops of Barritt's reagent A (α-naphthol) and 5 drops of Barritt's reagent B (KOH) to the tube.
5. Tap the bottom of the tube vigorously or shake it so that oxygen in the air aerates the medium. Take care that the culture does not spill out of

the tube. Place the tube in a rack for at least 30 minutes. The appearance of pink or red at the top of the broth is a positive result: the bacteria used the butanediol fermentation pathway. No change in color (yellow) is a negative result: the bacteria did not use this pathway.

6. Discard the tube in the designated place for sterilization.

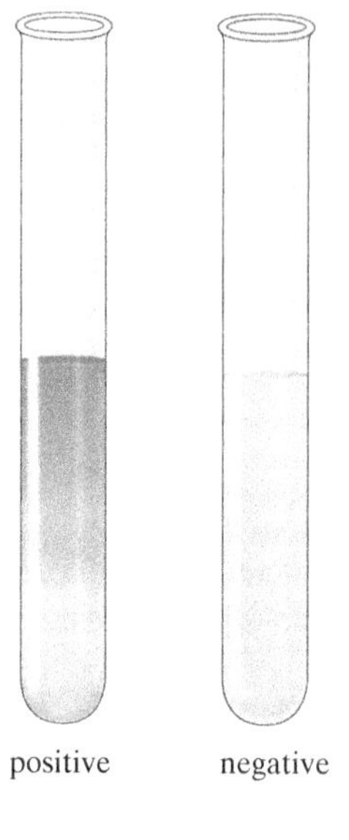

Interpretation of the Voges-Proskauer (butanediol fermentation) test

Troubleshooting

- If the tube is not aerated by vigorous shaking or tapping, the O_2 dissolved in the broth will be insufficient for the reaction of the reagents with acetoin. This will result in a false negative.
- Some bacteria need 48 hours of incubation for detectable levels of acetoin to accumulate in the medium.
- Addition of reagent A before reagent B will enhance a positive test.
- Do not use the same tube in which you did a methyl red test to do a Voges-Proskauer test. The reagents for each test will interfere with each other.

- *You must wear disposable gloves when you handle these reagents. Barritt's solution A (α-naphthol) is toxic and will penetrate the skin if it spills on you. Barritt's solution B (40% KOH) will burn unprotected skin.*
- *If you or anyone in the lab is allergic to latex (powder from latex gloves can elicit a severe allergic reaction in susceptible people), wear gloves made of nitrile.*
- *If any of the culture spills while you agitate the medium, properly disinfect the spilled mixture.*

UTILIZATION OF CARBOHYDRATES

Citrate Utilization Test

Purpose:

- To determine if a bacterium can utilize citrate as its sole source of carbon and energy.

Bacteria that can utilize citrate as the sole source of carbon and energy must have the membrane-associated transporter **citrate permease**. Once in the cell's cytoplasm, cellular enzymes convert citrate into pyruvate and CO_2.

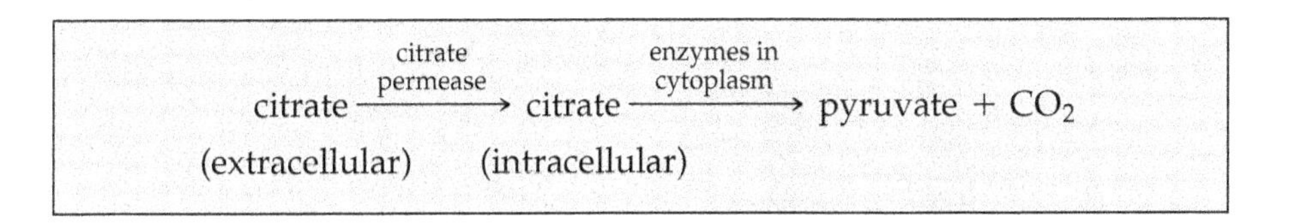

Pyruvate is used for energy. Other enzyme pathways construct biomolecules from pyruvate. Certain cellular enzymes build nitrogen-containing molecules using the NH_4^+ that is provided in the medium.

If the bacteria utilize citrate, the byproduct CO_2 combines with Na^+ in the medium to form $NaCO_3$, an alkaline compound. The pH indicator bromothymol blue turns blue in an alkaline pH. A change in color of the medium from green to blue denotes a positive test.

Materials:

- Simmons citrate agar slant tube
- pure culture of bacteria in a broth tube or on an agar slant

Procedure:

1. Label the Simmons citrate slant tube with the name of the bacterium you are testing, your name or initials, and the date.
2. Use aseptic technique to inoculate the Simmons citrate agar slant with either a loopful of bacteria from a broth culture or a small inoculum from an agar slant culture.
3. Incubate the inoculated slant at 35° C for up to 4 days.

4. Examine the slant each day for color change and for bacterial growth. A blue color is a positive test; the bacteria utilized citrate. A green color is a negative test.

5. Discard the tube in the designated place for sterilization.

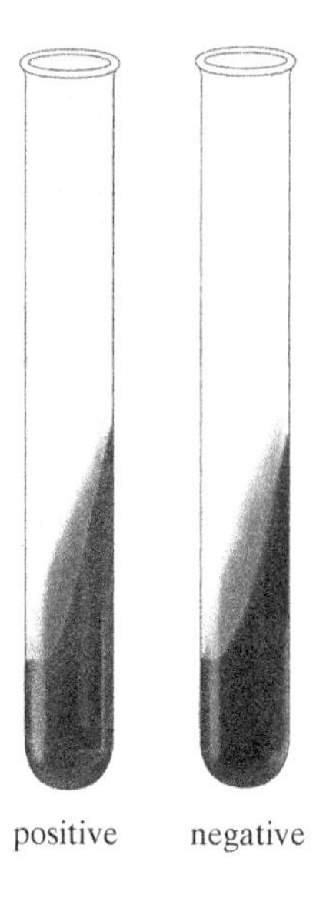

Interpretation of the citrate utilization test

Troubleshooting

- If you transfer a large inoculum to the agar slant, bacteria that cannot utilize citrate may still grow on the slant. This may result in a false positive.
- If you see a negative result after only 24 hours, the bacteria may simply utilize citrate at a slow rate. Allow the tube to incubate for up to 4 days.

Utilization of Carbohydrates

Oxidation–fermentation (OF) Glucose Test

Purpose:

- To determine if a gram-negative bacterium uses fermentation or aerobic respiration in utilizing a particular sugar for energy production.

To distinguish between bacteria that obtain their energy by fermentation or by aerobic respiration of sugars, an oxidation–fermentation (OF) medium contains a high concentration of a specific sugar and a low concentration of peptone (an enzymatic digest of protein). The small amount of peptone supports the growth of bacteria unable to utilize the sugar. To detect the production of organic acids during fermentation, the pH indicator bromthymol blue is included in the medium. If acids are present, the medium appears yellow: the bacteria have fermented the sugar in the tube. If no acids are present, the medium appears green: fermentation of the sugar did not take place.

Two tubes of OF medium are used to test a single bacterium: one tube is open to the air (aerobic conditions) and the other is covered with a layer of sterile mineral oil or melted paraffin to prevent diffusion of atmospheric oxygen into the medium (anaerobic conditions). Bacteria that ferment the sugar will produce organic acids in both tubes, and both tubes will appear yellow throughout. This result will also be seen for bacteria that have enzymatic pathways for both fermentation and aerobic respiration.

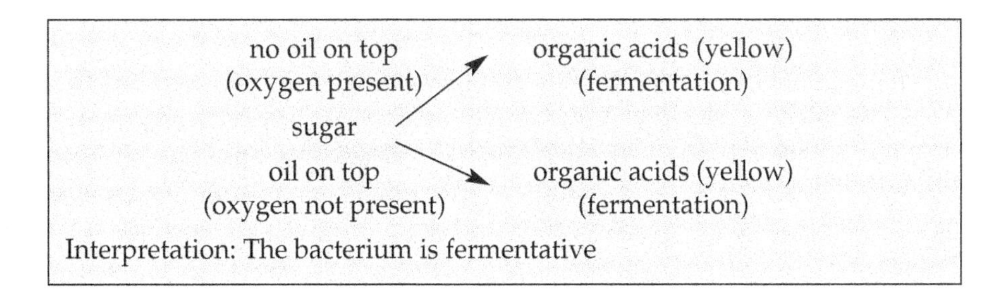

Bacteria that use an aerobic respiration pathway for extracting energy from the sugar will produce acid only in the tube open to the air. Only the medium in the unsealed tube will turn yellow at the top; the medium in the sealed tube will remain green.

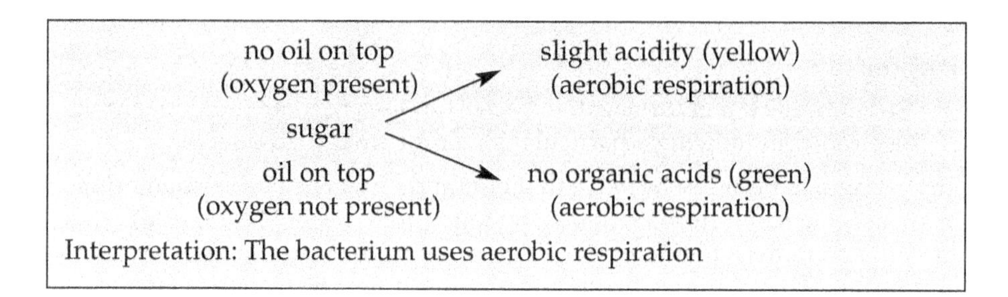

If the top of the medium in both tubes appears slightly yellow, it indicates either that the bacterium is only a slow fermenter, or is both an aerobic oxidizer and a slow fermenter.

If the color of the medium in both tubes remains green after incubation, the bacterium is considered to be a nonfermenter and a nonoxidizer of the sugar. If the bacterium lacks the enzyme to digest lactose, maltose, or sucrose, both OF tubes will also remain green after incubation.

An OF medium can also be used to detect motility because it contains semi-solid agar. Bacteria with flagella will swim away from the stab used to inoculate the medium, resulting in a "cloud" of bacterial growth around the stab.

Materials:

- 2 tubes with OF medium and a specific sugar (glucose, lactose, maltose, mannitol, sucrose, or xylose)
- sterile mineral oil or sterile melted paraffin with a sterile Pasteur pipette
- rubber bulb for pipette
- pure culture of bacteria in a broth tube or on an agar slant
- inoculating needle

Procedure:

1. Label the OF medium tube with the name of the bacterium you are testing, your name or initials, and the date.

2. Use a sterile needle to inoculate both of the OF medium tubes by stabbing three times with an inoculum taken from a broth culture or an agar slant culture.

3. Use aseptic technique to add 2–3 mm of sterile mineral oil or melted paraffin to the top of one of the tubes.

4. Incubate the inoculated tube at 35° C for at least 48 hours. After incubation, examine each tube for any change in color.

5. Discard the tubes in the designated place for sterilization.

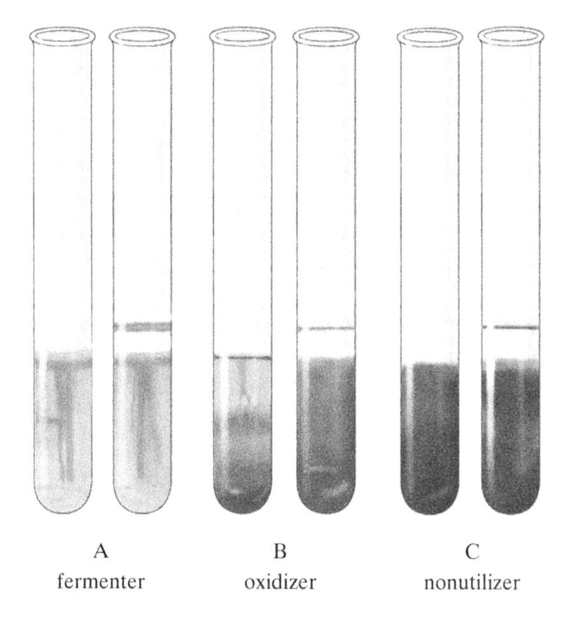

Interpretation of an oxidation–fermentation (OF) glucose test

Degradation of Amino Acids

Indole (Tryptophan Degradation) Test

Purpose:

- To determine the ability of some bacteria to split the amino acid tryptophan into indole and pyruvic acid.

Some bacteria can use tryptophan as a source of energy by degrading the amino acid to get pyruvate. **Indole** is a byproduct not used by the bacteria.

$$\text{tryptophan} + \tfrac{1}{2}\,O_2 \xrightarrow{\text{tryptophanase}} \text{pyruvate} + NH_3 + \text{indole}$$

Materials:

- tryptone broth tube or SIM (sulfide-indole-motility) agar deep
- pure culture of bacteria in a broth culture or on an agar slant
- Kovac's reagent in a dropper bottle or in a bottle with a Pasteur pipette and a bulb
- disposable gloves

Procedure:

1. Label the tube with either tryptone broth or SIM agar deep with the name of bacterium you are testing, your name or initials, and the date.
2. Use aseptic technique to inoculate the tryptone broth tube with either a loopful of bacteria from a broth culture or a small inoculum from an agar slant culture. For an SIM agar deep tube, use an inoculating needle to stab the medium with an inoculum from either a broth culture or an agar slant culture.
3. Incubate the inoculated tube at 35° C for 24–48 hours.
4. After incubation, add 5 drops of Kovac's reagent to the culture.
5. The quick appearance of a red layer at the top of the tube is a positive test for the presence of indole. The absence of a red layer is a negative test: tryptophan was not hydrolyzed.

6. Discard the tube in the designated place for sterilization.

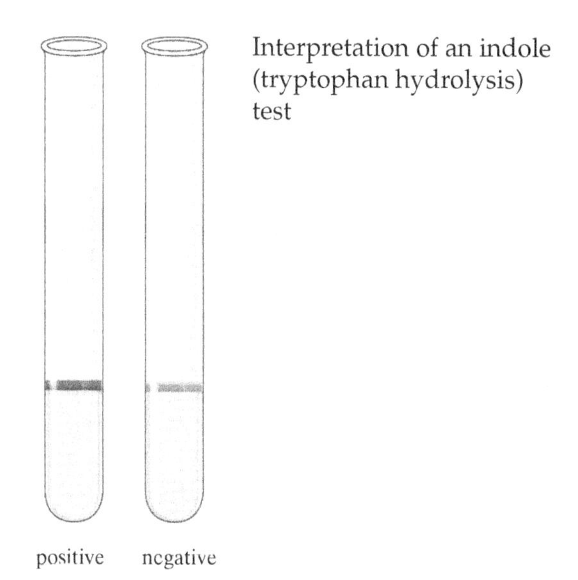

Interpretation of an indole (tryptophan hydrolysis) test

- *You must wear disposable gloves when you handle Kovac's reagent. This reagent contains the harmful chemicals amyl alcohol, concentrated HCl, and p-(dimethylamino)benzaldehyde.*
- *If you or anyone in the lab is allergic to latex (powder from latex gloves can elicit a severe allergic reaction in susceptible people), wear gloves made of nitrile.*

DEGRADATION OF AMINO ACIDS

Hydrogen Sulfide (H_2S) Production Test

Purpose:

- To determine if a bacterium is able to dismantle the amino acid cysteine or to reduce thiosulfate and form hydrogen sulfide as a byproduct.

Some bacteria use the enzyme **cysteine desulfurase** to break down cysteine, a sulfur-containing amino acid, to form pyruvate, which is then utilized for energy. **Hydrogen sulfide (H_2S)** is a byproduct not used by the bacteria. Other bacteria reduce thiosulfate in energy-generating pathways and also release H_2S.

To detect H_2S production, ferrous (Fe^{+2}) ions are included in growth media that contain protein and thiosulfate. Fe^{+2} combines with S^{-2} to form a black precipitate (FeS).

Three media can be used to detect H_2S−producing bacteria: peptone iron agar, SIM (sulfide-indole-motility) agar, and triple sugar iron (TSI) agar. SIM agar also detects indole-producing and motile bacteria. TSI agar is explained in another section.

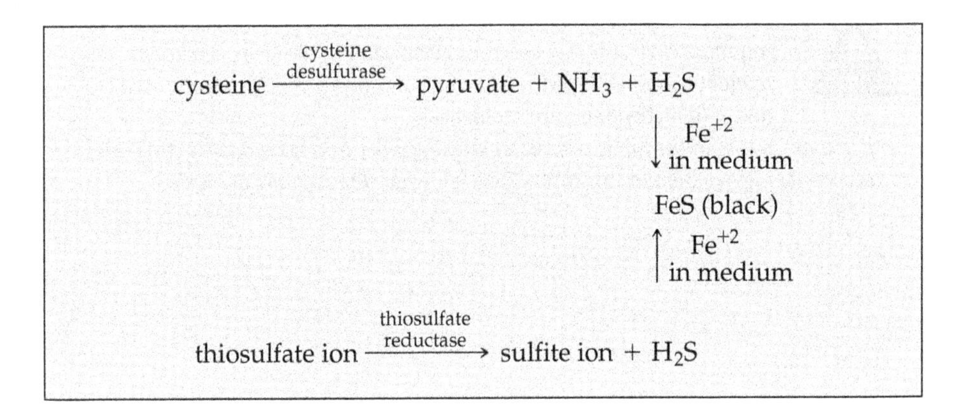

Materials:

- Peptone iron agar deep tube or SIM agar deep tube
- pure culture of bacteria on an agar slant or agar plate

Procedure:

1. Label the tube of medium with the name of the bacterium you are testing, your name or initials, and the date.
2. Use a sterile inoculating needle to pick up a small inoculum of the bacterial culture. Inoculate the medium by a single stab into the agar to three-fourths depth.
3. Incubate the inoculated tube at 35° C for up to 7 days. Examine the tube initially at 24–48 hours.
4. Examine the tube for blackening in the agar.
5. If you have used SIM agar, you can also check for bacterial motility by observing that the black precipitate extends beyond the stab.
6. Discard the tube in the proper place for sterilization.

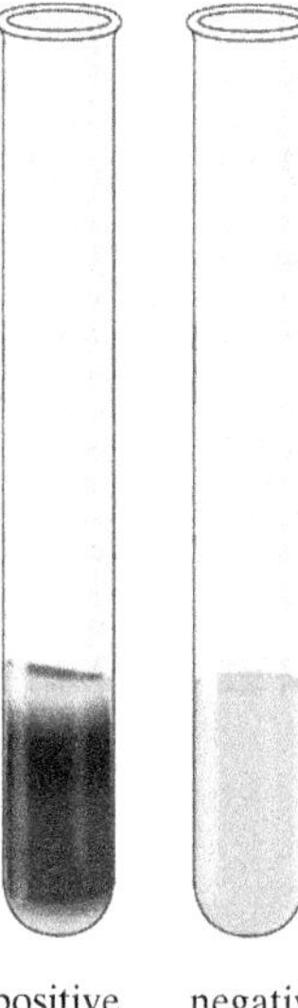

Interpretation of an H_2S production test

DEGRADATION OF AMINO ACIDS

Phenylalanine Deamination Test

Purpose:

- To determine the ability of a bacterium to remove the amino group (deaminate) from the amino acid phenylalanine.

Members of the genera *Proteus* and *Providencia* use the enzyme **phenylalanine deaminase** to catalyze the removal of the amino group from phenylalanine. In addition to NH_3, phenylpyruvic acid is the major organic product. The presence of phenylpyruvic acid is detected by the addition of a solution of ferric chloride ($FeCl_3$). A dark green color appears in a positive test. If no phenylpyruvic acid is present, the medium appears yellow, the color of $FeCl_3$.

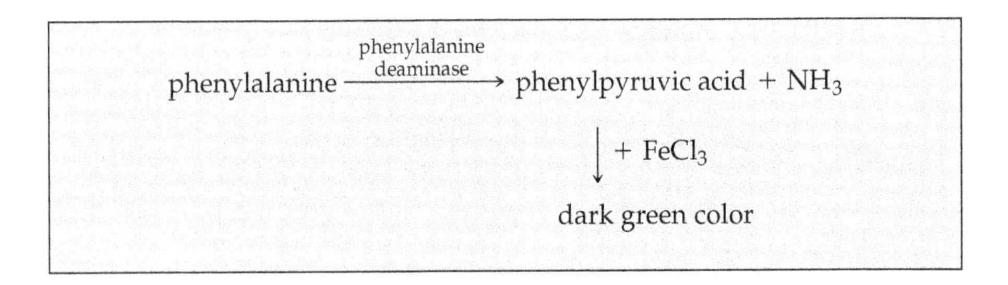

Materials:

- phenylalanine agar slant tube
- 10% $FeCl_3$ solution in a dropper bottle or in a bottle with a Pasteur pipette and a bulb
- pure culture of bacteria in a broth medium or on an agar slant

Procedure:

1. Label the phenylalanine agar slant tube with the name of the bacterium you are testing, your name or initials, and the date.
2. Use a sterile inoculating loop to transfer bacteria from either a broth culture or an agar slant onto the surface of the phenylalanine agar slant.
3. Incubate the inoculated tube at 35° C for 18–24 hours.
4. After incubation, add 3–5 drops of 10% $FeCl_3$ to the slant. Tilt and turn the tube so that the reagent covers the slant surface.

5. Examine the tube after 1–5 minutes for the appearance of a dark green color. This denotes a positive test. If you see no color change after 5 minutes, the test is negative.

6. Discard the tube in the designated place for sterilization.

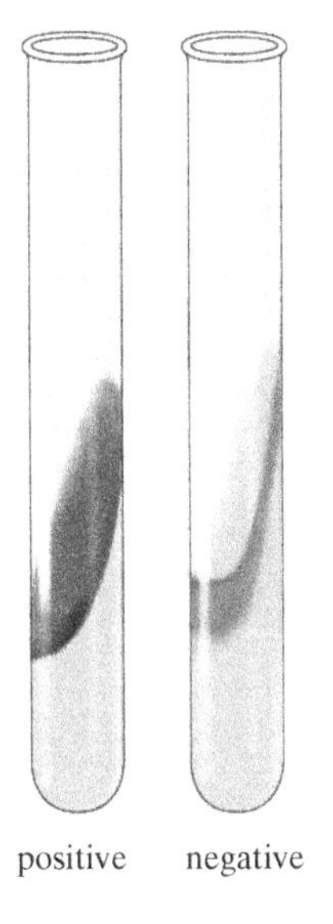

Interpretation of a phenylalanine deamination test

Troubleshooting

- Because the dark green color fades, read the results no later than 5 minutes after you added the drops of the $FeCl_3$ solution.
- If you do not allow the $FeCl_3$ solution to cover the agar slant surface, an insufficient amount of the reagent will diffuse through the agar to react with phenylpyruvic acid.

Degradation of Amino Acids

Amino Acid Decarboxylase Test

Purpose:

- To determine the ability of a bacterium to remove the carboxyl group from an amino acid (decarboxylation).

Members of the bacterial family *Enterobacteriaceae* can be distinguished by their production of a specific **decarboxylase** that removes the carboxyl group from particular amino acids. These amino acids include the amino acids lysine, ornithine, and arginine.

The reaction is assessed by changes in the color of the pH indicator in the medium. Fermentation of glucose in the medium results in accumulation of acids (purple to yellow). However, with decarboxylation of the amino acid, the resulting amine causes an increase in pH (yellow to purple). A positive result for an amino acid decarboxylase is purple in the medium; a negative result is yellow.

$$\text{amino acid} \xrightarrow{\text{decarboxylase}} \text{amine} + CO_2$$

(lysine) → (cadaverine)
(ornithine) → (putrescine)

Materials:

- decarboxylase medium broth tube with lysine, ornithine, or arginine
- sterile mineral oil
- sterile Pasteur pipette and rubber bulb
- pure culture of bacteria in broth culture or on an agar slant

Procedure:

1. Label the amino acid–decarboxylase medium broth tube with the amino acid, the name of bacterium you are testing, your name or initials, and the date. Note that the medium is purple.

2. Use aseptic technique to inoculate the decarboxylase broth tube with either a loopful of bacteria from a broth culture or a small inoculum from an agar slant culture.

3. Use a sterile Pasteur pipette and bulb to aseptically add a 4 mm layer of sterile mineral oil to the top of the inoculated medium.

4. Incubate the inoculated tube at 35° C for up to 4 days. Check the tube for a color change every 24 hours. A yellow color indicates an acid pH due to accumulation of organic acids from fermentation of glucose. The appearance of purple after yellow denotes a positive test for decarboxylase. If yellow continues after several days of incubation, the test is negative.

5. Discard the tube in the designated place for sterilization.

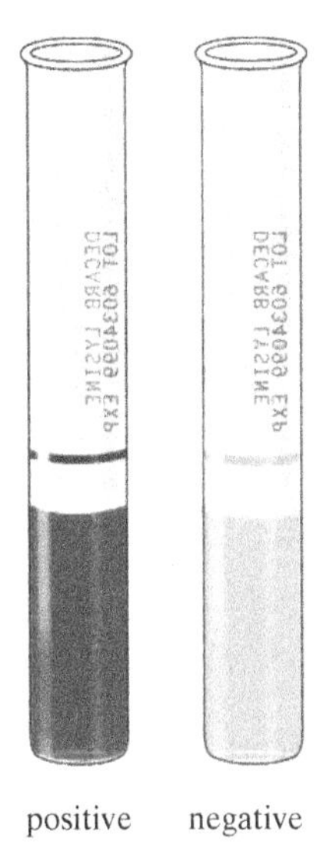

positive negative

Interpretation of a decarboxylase test (lysine)

Troubleshooting

- If you omit adding the layer of sterile mineral oil, air can cause a false increase in pH at the surface. This can create a false positive test if the bacterium is decarboxylase-negative.
- If both purple and yellow are present after incubation, gently shake the tube to mix the contents.

RESPIRATION TESTS
Catalase Test

Purpose:

- To detect the presence of catalase, an enzyme that degrades hydrogen peroxide.

In aerobic respiration, hydrogen peroxide (H_2O_2) is a reactive intermediate that forms as oxygen is reduced. H_2O_2 damages DNA and cell membranes and alters the active sites of some enzymes. Bacteria that tolerate oxygen (O_2) or require the gas for metabolism use the enzyme **catalase** to quickly break down H_2O_2 into water and O_2.

$$2\,H_2O_2 \xrightarrow{\text{catalase}} 2\,H_2O + O_2\,(\text{gas})$$

When drops of a dilute solution of H_2O_2 are added onto bacteria containing catalase, bubbles of O_2 rapidly appear. These bubbles denote a positive test.

Materials:

- 3% H_2O_2 solution
- Pasteur pipette and a bulb
- pure culture of bacteria on an agar surface (agar slant or agar plate)
- sterile wood stick
- clean microscope slide

Procedure:

1. Use a Pasteur pipette to add a few drops of 3% H_2O_2 solution to one of the following:
   - a few bacterial colonies growing on an agar plate
   - bacteria growing on an agar slant
   - a small amount of bacteria that has been transferred from an agar surface with a sterile wooden stick onto the surface of a clean microscope slide. Dispose of the stick in a biohazard bag.

2. The quick appearance of vigorous bubbling is a positive test for catalase. No or very weak bubbling is a negative test.

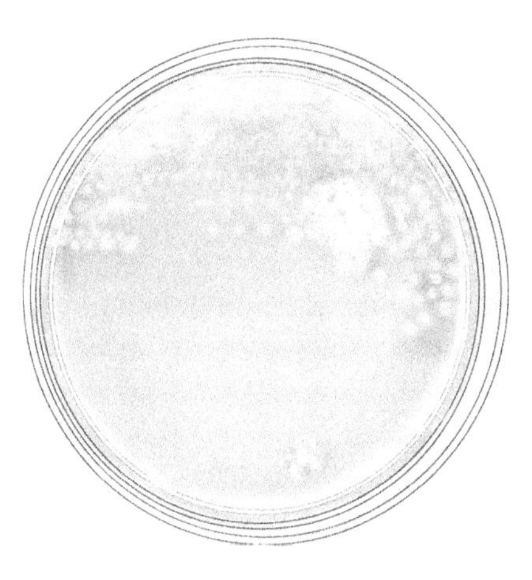

Positive catalase test (note bubbling)

Troubleshooting

- The metal in an inoculating loop used to transfer bacteria to the slide may non-specifically degrade H_2O_2, giving a false positive result.
- A false positive result can follow testing bacteria that are growing on a blood agar plate. Red blood cells, which contains catalase, may contaminate the inoculum.

RESPIRATION TESTS

Oxidase Test

Purpose:

- To determine if bacteria have cytochrome oxidase, a participant in electron transport during respiration.

Electron transport molecules in bacterial plasma membranes use the energy in electrons removed from food molecules to create a proton motive force across the membrane. This energy is then converted into forms that the cell can use for work. At the end of the electron transport chain, **cytochrome (*c*) oxidase** collects electrons and facilitates their addition to molecular O_2 and with H^+ to form H_2O. Some bacteria use cytochrome oxidase to add electrons to nitrate when O_2 is not available (nitrate reduction). Not all bacteria that can grow in O_2 have cytochrome (*c*) oxidase.

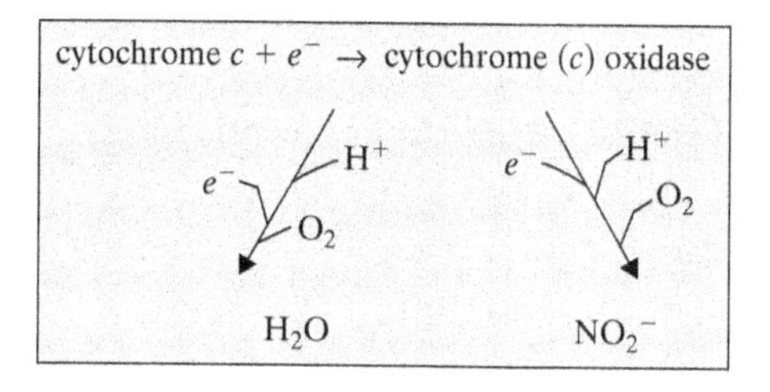

When cytochrome oxidase adds electrons to the oxidase reagent, the reduced form turns dark blue to purple within seconds.

Materials:

- oxidase reagent – N,N,N,N tetramethyl-p-phenylenediamine: available in a crushable ampule or impregnated in paper
- sterile wood stick
- piece of filter paper (Whatman No. 1 or 2)
- pure culture of bacteria on an agar surface (agar slant or agar plate)

Procedure:

1. If you are using a crushable ampule, hold it upright and pointing away from you. With your forefinger and thumb in the middle, squeeze to break the glass inside the plastic dropper. Tap the bottom of the ampule on the bench top. Then add a few drops of the oxidase reagent to one of the following:
    - a bacterial colony growing on an agar plate
    - a small amount bacteria that has been transferred from on an agar slant with a sterile wooden stick onto a piece of filter paper. Dispose of the stick in a biohazard bag.

2. If within 10 seconds the bacteria turn deep blue to purple, the reaction is positive for oxidase. Dispose of the filter paper in a biohazard bag.
3. If the oxidase reagent is available already impregnated in a paper strip, use a sterile wood stick to transfer bacteria from an agar surface onto the paper strip. Add a few drops of sterile water to the bacteria on the strip. Dispose of the wood stick in a biohazard bag. Observe as above for the rapid appearance of purple.

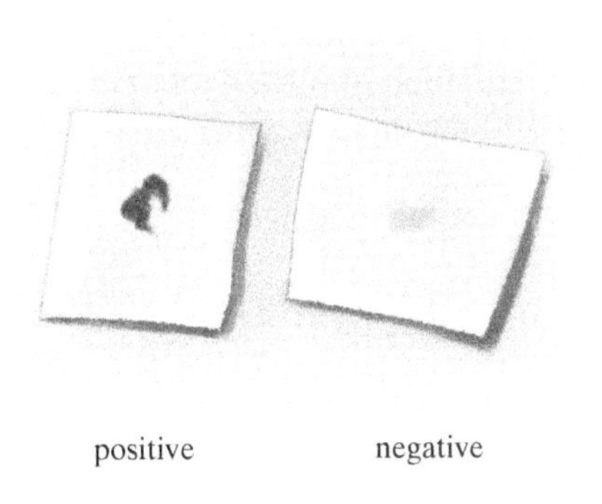

positive negative

Interpretation of an oxidase test on filter paper

Troubleshooting

- If your inoculating loop is made of nichrome wire, iron in the metal alloy will cause a false positive reaction. The iron will catalyze an unwanted oxidation of the reagent
- Cultures must be no more than 24 hours old. They also must be at room temperature.
- Use only fresh reagent. Oxygen in the air can affect the reagent and turn it purple.
- If the purple color develops *after* 10 seconds, the reaction is negative. Oxygen in the air may have oxidized the reagent.

- *If you used a wood stick to transfer bacteria, place the stick in a biohazard bag or in a beaker with disinfectant.*
- *If you used the filter paper method or the reagent/paper strip, place the paper with bacteria and reagent in a biohazard bag or other receptacle for sterilization.*

Respiration Tests

Nitrate Reduction Test

Purpose:

- To determine if a bacterium is able to reduce nitrate ions to either nitrite ions or to nitrogen gas.

In anaerobic respiration, bacteria add electrons that have been passed along the electron transport chain to an inorganic substance that is not oxygen. One such final electron acceptor is the nitrate ion. The enzyme complex **nitrate reductase** facilitates this reduction of nitrate ion. For some bacteria that do this, nitrate ions are reduced to nitrite ions:

$$\underset{\text{(nitrate ion)}}{NO_3^- + 2\,H^+ + 2\,e^-} \xrightarrow{\text{nitrate reductase}} \underset{\text{(nitrite ion)}}{NO_2^- + H_2O}$$

Other bacteria that use nitrate reduction during anaerobic conditions for energy production are able to reduce nitrate completely to molecular nitrogen. This is called **denitrification**.

$$2\,NO_3^- + 12\,H^+ + 10\,e^- \xrightarrow{\text{nitrate reductase}} 2\,NO_2^- \longrightarrow \longrightarrow N_2 + 6\,H_2O$$

To detect nitrite ions that are products of nitrate reduction by bacteria inoculated in nitrate broth, two reagents are added: sulfanilic acid (reagent A) and dimethyl-α-naphthylamine (reagent B). If nitrite is present, the medium turns pink or red. This is a positive test for nitrate reduction.

But the absence of a color change cannot be interpreted as a negative test. Either the nitrate ions were not reduced to nitrite ions (a true negative test) or the nitrate ions were completely reduced to molecular nitrogen (a positive test). To distinguish between these two possibilities, a small amount of zinc is added to the tube, which already contains nitrate reagents A and B. Zinc will reduce nitrate ions to nitrite ions. If the bacteria did not reduce nitrate ions, zinc will do this. Therefore, the appearance of a pink or red color after addition of zinc is interpreted as a negative result. If, however, no color change is observed after zinc is added, the test is positive for nitrate reduction. No nitrate ions remain in the broth because they were completely reduced to molecular nitrogen.

Materials:

- nitrate broth tube
- nitrate reagent A (sulfanilic acid) in a dropper bottle or in a bottle with a Pasteur pipette and a rubber bulb
- nitrate reagent B (dimethyl-α-naphthylamine) in a dropper bottle or in a bottle with a Pasteur pipette and a rubber bulb
- powdered zinc
- a small spatula or a wooden stick for adding zinc powder
- disposable gloves
- pure culture of bacteria on an agar slant

Procedure:

1. Label the nitrate broth tube with the name of the bacterium you are testing, your name or initials, and the date.
2. Use aseptic technique to inoculate the nitrate broth tube with a small inoculum from an agar slant culture.
3. Incubate the inoculated tube at 35° C for 24–48 hours.
4. After incubation, add 5 drops of reagent A and 5 drops of reagent B to the tube. Gently shake the tube to mix the reagents with the broth. *Wear eye protection and disposable gloves when you add the nitrate reagents.*
5. If a pink to red color develops in 1–2 minutes, the test is positive for nitrate reduction (nitrite ions are present).
6. If no color change is seen, add a small amount of powdered zinc to the tube. If you do not observe a color change after 10 minutes, the test is positive for nitrate reduction. If you observe a color change to pink or red, the test is negative.
7. Discard the tube in the designated place for sterilization.

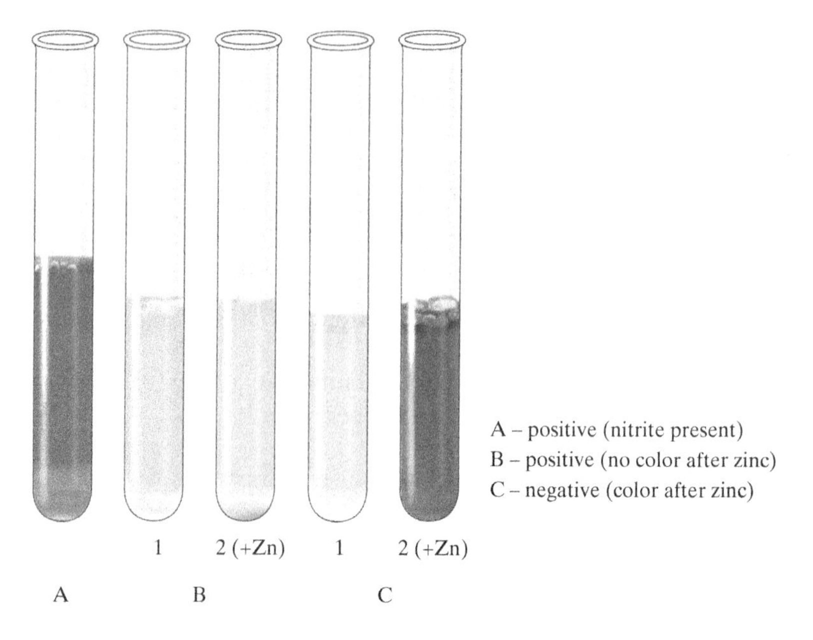

Interpretation of a nitrate reduction test

- *Wear eye protection and disposable gloves when you add reagents A and B Reagent A (sulfanilic acid) is caustic. Reagent B (dimethyl-α-naphthy lamine) is a possible carcinogen. Wash immediately with lots of water if you spill either reagent on your skin or eyes. If either reagent gets into your eyes, seek immediate medical assistance after you wash your eyes.*
- *If you or anyone in the lab is allergic to latex (powder from latex gloves can elicit a severe allergic reaction in susceptible people), wear gloves made of nitrile.*
- *Take care that you do not inhale the powdered zinc and that it does not contact your skin. If you spill any zinc on your skin, wash it off immediately with lots of water.*
- *Keep open flames away from powdered zinc. A cloud of powder will explode if it gets too close to a flame.*

Troubleshooting

- If you add too much zinc powder, you may get a false negative result or a short-lived color change.
- Because the assay for nitrite is very sensitive, you might add reagents A and B to an uninoculated tube of nitrate broth to verify that the medium contains no nitrite.

MISCELLANEOUS TESTS

Triple Sugar Iron (TSI) Agar Test

Purpose:

- To differentiate among the gram-negative enteric bacilli (*Enterobacteriaceae*) as to their ability to ferment glucose, lactose, and sucrose and to produce H_2S.

Triple sugar iron (TSI) agar contains 1% lactose and sucrose and 0.1% glucose (dextrose). If a bacterium ferments any of these sugars, the organic acids that are waste products cause a drop in pH. The small amount of glucose is rapidly consumed if the bacterium ferments it; the bacterium then digests the protein in the medium, liberating ammonia from the amino acids. This raises the pH to become alkaline. Phenol red is present as the pH indicator, appearing yellow in acid pH and red in alkaline pH.

If a bacterium can reduce thiosulfate, also included in the medium, hydrogen sulfide (H_2S) is produced. Ferrous ions (Fe^{+2}) in the medium will combine with sulfide ions (S^{-2}) and FeS appears as a black precipitate. H_2S may also be a byproduct of the dismantling of the amino acid cysteine in the protein also present in the medium.

TSI agar is prepared as a slant. An inoculating needle is used to stab the agar deep portion (the "butt") and to streak the slant in a zigzag pattern. Several observations are possible after incubation of the inoculated tube. These reactions and their interpretation appear in the table below.

Observation	Interpretation
Yellow slant/yellow butt	Glucose and lactose and/or sucrose were fermented
Red slant/yellow butt	Only glucose was fermented
Red slant/red butt	None of the sugars were fermented Not a member of the *Enterobacteriaceae*
Cracks or bubbles in yellow agar	Gas was a byproduct of fermentation
Black precipitate in butt	H_2S was produced

It is important that you observe the tube for any color change between 18–24 hours after you incubate the tube.

Materials:

- triple sugar iron (TSI) agar slant tube
- pure culture of a bacterium growing on an agar slant
- inoculating needle with a straight wire

Procedure:

1. Label the TSI tube with the name of bacterium you are testing, your name or initials, and the date.
2. Use aseptic technique to remove a small inoculum of bacteria from an agar slant surface with an inoculating needle. Stab the bacteria into the TSI agar butt to about three-quarters depth. As you withdraw the needle, streak in a zigzag pattern along the slant surface. If the tube has a screw-on cap, do not completely tighten it; allow for air to enter the tube.
3. Incubate the inoculated tube at 35° C for 18–24 hours.
4. After incubation, examine the tube for any changes in the color of the butt and slant, for the appearance of gas, and for a black precipitate.
5. Discard the tube in the designated place for sterilization.

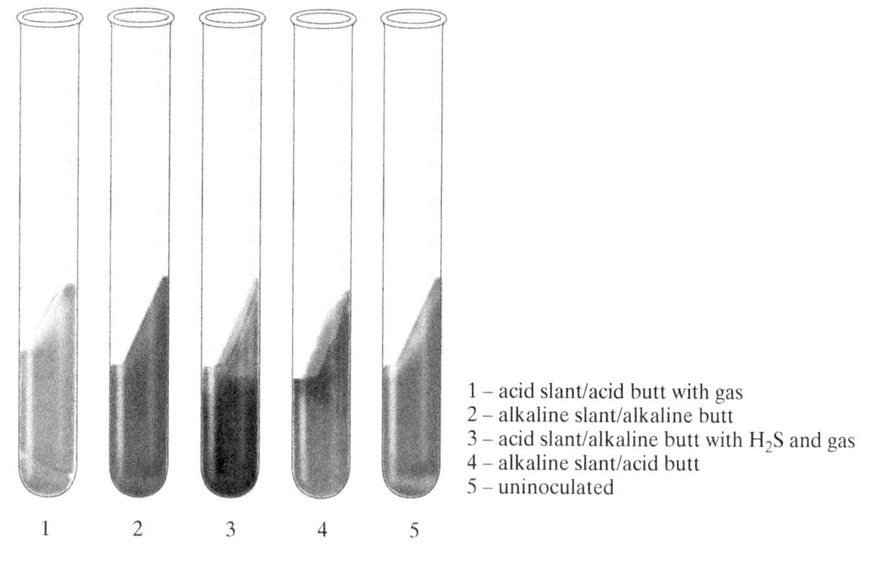

Interpretation of results for TSI agar slant tests

Troubleshooting

- If you use an inoculating loop instead of a needle to stab the butt, the agar will crack. The outcome will be a false positive for gas production.
- If the cap on a screw-cap tube is completely tightened, the lack of air will reduce possible alkaline responses.
- If a black precipitate in the butt masks its color, consider the color to be yellow, indicating an acid pH.

MISCELLANEOUS TESTS

Urea Hydrolysis Test

Purpose:

- To determine the ability of a bacterium to hydrolyze urea.

Urea is a small molecule excreted in urine to eliminate excess nitrogen from the body. A few bacteria use the enzyme **urease** to rapidly degrade urea into carbon dioxide and ammonia.

$$\text{urea} \xrightarrow{\text{urease}} \text{carbon dioxide} + \underset{\text{(raises pH)}}{\text{ammonia}}$$

As urease-producing bacteria grow in a medium containing urea, ammonia accumulates, making the medium more alkaline. This pH change is detected with phenol red, which turns bright pink in alkaline pH. A positive test for urease, therefore, is the appearance of bright pink in the medium.

Materials:

- urea-containing medium: broth or agar slant
- pure culture of bacteria in a broth tube or on an agar slant

Procedure:

1. Label the tube of urea medium with the name of the bacterium you are testing, your name or initials, and the date.
2. Use aseptic technique to inoculate the urea medium tube with either a loopful of bacteria from a broth culture or a small inoculum from an agar slant culture.
3. Incubate the inoculated tube at 35° C for no more than 24 hours.
4. Examine the tube for any color change. The appearance of a bright pink color is a positive test for urease. No change in the color of the medium is a negative test.

5. Discard the tube in the designated place for sterilization.

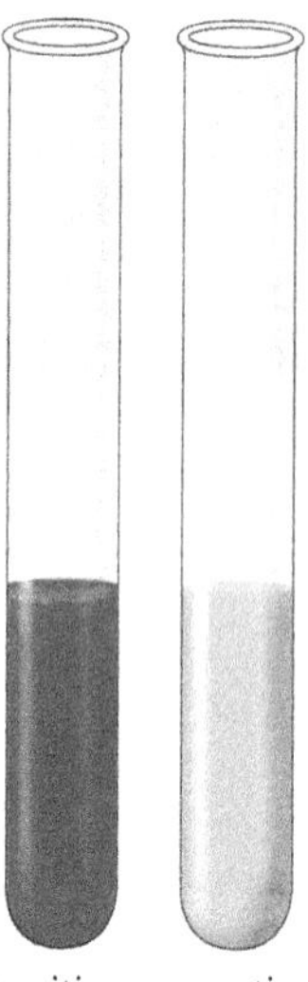

Interpretation of a urea hydrolysis test

MISCELLANEOUS TESTS

Litmus Milk Reactions

Purpose:

- To differentiate among bacteria as to their ability to utilize lactose, protein, and litmus in litmus milk.

Bacteria differ in their ability to metabolize different substrates in milk, including lactose and milk proteins. These differences can be determined in litmus milk. Litmus serves as a pH indicator, turning pink in acid pH and blue in alkaline pH. Its color in an uninoculated medium is lavender (purplish-blue). Litmus can also be reduced (accept electrons) by certain bacterial enzymes.

Bacteria that ferment lactose form organic acids. The resulting acid pH turns litmus pink.

$$\underset{\text{(lavender medium)}}{\text{lactose}} \xrightarrow{\text{fermentation}} \underset{\text{(pink medium)}}{\text{acids (pH} < 4.5)}$$

If sufficient acid is produced, casein (milk protein) will become denatured and form into a firm curd (acid curd) that remains in place when the tube is tilted. Whey is the clear, watery fluid at the top of the curd.

$$\begin{array}{l} \text{lactose} \xrightarrow{\text{fermentation}} \text{acid (pH} < 4.5) \\ \quad\downarrow \\ \text{casein} \longrightarrow \text{firm (acid) curd} \end{array}$$

Some bacteria that form an acid curd in litmus milk will also reduce litmus to a colorless compound. Where litmus is reduced, the acid curd will appear white, beginning at the bottom of the tube.

$$\underset{\text{(pink)}}{\text{litmus (acid pH)}} \xrightarrow{+\,e^{-}\text{ from food}} \underset{\text{(colorless)}}{\text{litmus (reduced)}}$$

A soft curd is formed when casein is altered by the bacterial enzyme rennin. This curd is "runny" when the tube is tilted. Because the pH does not change, the color of the medium does not change.

$$\text{casein} \xrightarrow{\text{rennin}} \text{soft curd}$$

Proteolytic bacteria digest milk proteins into peptides and amino acids. (sometimes called peptonization). The medium will clear when a curd or insoluble casein is completely digested.

$$\underset{\text{(insoluble suspension)}}{\text{casein (curd)}} \xrightarrow{\text{proteolytic enzymes}} \underset{\text{(clearing)}}{\text{peptides + amino acids}}$$

In the first 24 hours after inoculation, proteolytic bacteria may cause an alkaline pH in litmus milk. Other bacterial enzymes dismantle some amino acids, releasing ammonia. Litmus turns blue in the alkaline pH.

$$\underset{\text{(lavender)}}{\text{casein}} \xrightarrow{\text{proteases}} \text{amino acids} \xrightarrow{\text{deaminases}} \underset{\text{(blue)}}{NH_3 \text{ (pH} \uparrow)}$$

Materials:

- litmus milk tube
- pure culture of bacteria in a broth tube or on an agar slant

Procedure:

1. Label the litmus milk tube with the name of the bacterium you are testing, your name or initials, and the date.
2. Use aseptic technique to inoculate the litmus milk tube with either a loopful of bacteria from a broth culture or a small inoculum from an agar slant culture.

3. Incubate the inoculated tube at 35° C for up to 7 days. Examine the tube for its litmus milk reaction each day.
4. Discard the tube in the designated place for sterilization.

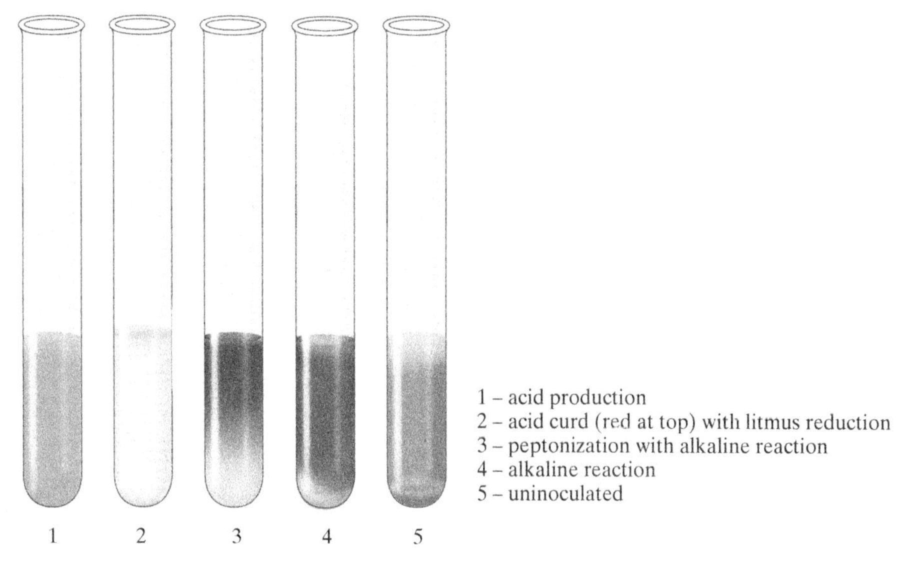

Interpretation of litmus milk reactions

Troubleshooting

- When you prepare litmus milk medium, a final pH of 6.8 ± 0.2 is essential so that the pH changes due to bacterial metabolism can be detected by color change in litmus. The uninoculated medium should appear lavender or purplish-blue.

MISCELLANEOUS TESTS

Motility Testing

Purpose:

- To determine if a bacterium is motile.

Bacterial motility is most often due to one or more flagella that propel the cell by spinning like a corkscrew. When motile bacteria are stabbed with an inoculating needle into semisolid agar (0.4% agar rather than 1.5%), the cells swim away from the stab. A diffuse "cloud" of growth extending from the stab indicates motility. Non-motile bacteria will grow only along the stab.

You can also detect motility using a hanging-drop slide preparation, (described on pp. 84–86) but this method is not suitable for pathogenic bacteria because of the risk of environmental contamination.

Materials:

- a tube with motility test medium or SIM (sulfide-indole-motility) agar
- an inoculating needle with a straight wire
- pure culture of bacteria in a broth medium or on an agar surface

Procedure:

1. Label the tube with the name of bacterium you are testing, your name or initials, and the date.
2. Use aseptic technique to dip a sterile inoculating needle into a broth culture or to remove a small inoculum from an agar surface. Then stab the bacteria into the center of the test medium to about halfway to three-quarters depth.
3. Incubate the inoculated tube at 35° C for 24–48 hours.
4. After incubation, examine the tube for bacterial growth that spreads out from the stab. Non-motile bacteria will grow only along the stab. If you do not observe motility after 48 hours, continue to incubate for up to 7 days.
5. Discard the tube in the designated place for sterilization.

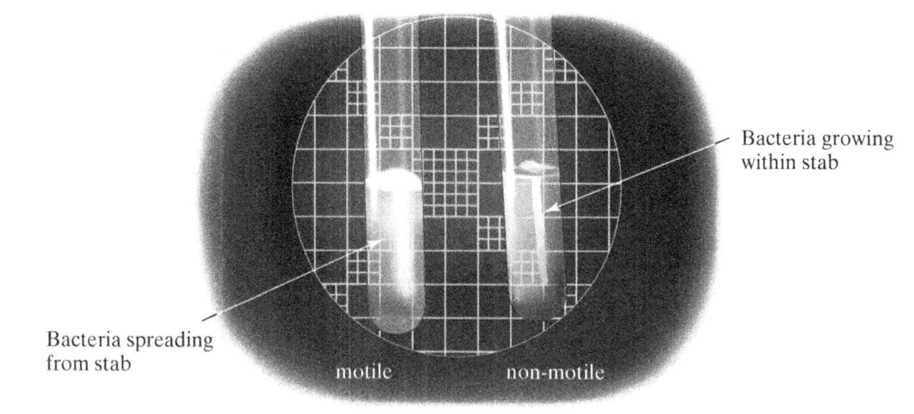

Appearance of motility test medium

MISCELLANEOUS TESTS

Coagulase (Tube) Test

Purpose:

- To distinguish between pathogenic and non-pathogenic staphylococci, based on their ability to cause blood plasma to clot with the enzyme coagulase.

Coagulase is an enzyme made by *Staphylococcus aureus* that activates a pathway that converts fibrinogen in blood plasma into fibrin, the protein threads that stick together to form a clot. *S. aureus* makes two forms of coagulase, one extracellular or **free** and the other **bound** to the cell wall. The tube test detects both free and bound coagulase in *S. aureus*. In the test, an inoculum of bacteria is mixed with rabbit plasma in a small tube and incubated at 35° C. The appearance of a solid gelled clot is a positive reaction for coagulase. During preparation of the plasma by the manufacturer, citrate or EDTA are added to bind Ca^{+2} to block unwanted clotting.

Materials:

- rabbit plasma (with citrate or EDTA) reconstituted with sterile water
- sterile 1.0 mL pipette and pipetter *or* 1.0 mL syringe with needle
- small test tube (10 × 75 mm)
- water bath set at 35° C *or* 35° C incubator
- sterile Pasteur pipette with rubber bulb (optional)
- pure culture of bacteria grown in an overnight broth culture or on an agar surface
- disposable gloves

Procedure:

1. Use aseptic technique to add 0.5 mL of rabbit plasma to a small test tube with either a sterile 1.0 mL pipette or a sterile 1.0 mL syringe with needle.
2. Label the tube containing rabbit plasma with the source of the specimen or the name of the bacterium you are testing, your name or initials, and the time and date.
3. Use a sterile Pasteur pipette to transfer 2–3 drops of broth culture to the tube with rabbit plasma.

   Or

   Use a sterile inoculating loop to transfer a heavy inoculum of bacteria from an agar surface to the tube with rabbit plasma.

4. Incubate the tube at 35° C for 4 hours.
5. After incubation, examine the tube for the appearance of a gel-like clot. If the tube is negative for clot formation, incubate the tube at room temperature for an additional 20 hours.
6. When you have completed your examination of the tube, place it in the designated place for sterilization.

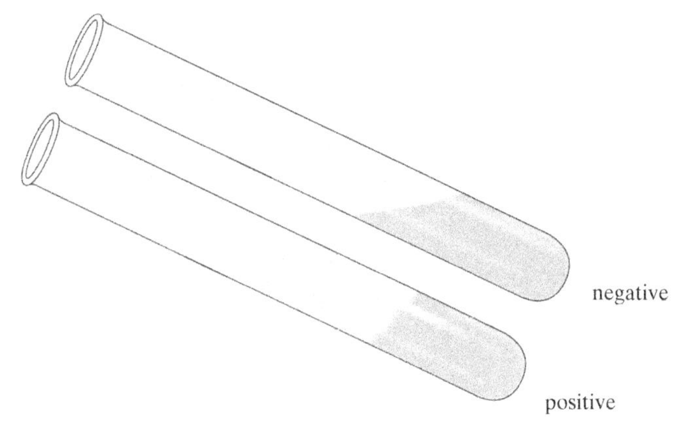

Appearance of a coagulase tube test

- *If you are testing S. aureus, a potential pathogen, you must wear disposable gloves. If you or anyone in the lab is allergic to latex (powder from latex gloves can elicit a severe allergic reaction in susceptible people), wear gloves made of nitrile.*
- *Spilled material that contains S. aureus must be promptly disinfected.*

Selective and/or Differential Media

Blood Agar Plate

Purpose:

- To isolate and support the growth of fastidious bacteria.
- To differentiate among bacteria based on their ability to lyse red blood cells (hemolysis).

Blood agar is an **enriched** medium prepared with 5% sheep red blood cells (RBCs). This nutritionally abundant medium is used to isolate and grow fastidious bacteria in clinical specimens, including those taken from skin, nose, throat, cerebrospinal fluid, and blood. Blood agar is also **differential** because it distinguishes among bacteria as to their ability to lyse RBCs. In **β-hemolysis**, bacteria secrete enzymes that completely dismantle the RBCs. A clear zone appears around the bacterial colony. In **α-hemolysis**, partial disruption of the RBCs by the bacteria results in a greenish hue around a colony. Nonhemolytic bacteria do not damage RBCs and are sometimes called γ-hemolytic.

Materials:

- blood agar plate
- a specimen that contains a mixture of bacteria

  or

  a pure culture of a bacterium in a broth medium
- candle jar

Procedure:

1. Label the blood agar plate with the source of the specimen or the name of the bacterium you are testing, your name or initials, and the date.
2. Use aseptic technique to inoculate the blood agar plate as a streak plate (pp. 43–45). This will isolate individual colonies from a specimen of mixed bacteria.

2a. If you are testing for the hemolytic activity of a bacterium growing on blood agar, use aseptic technique to prepare a streak plate with a loopful of broth culture.

3. Incubate the inoculated plate upside-down for 24–48 hours at 35° C.

4. Examine the plate for a hemolytic reaction. For an accurate reading, observe the plate lit from behind.
5. Discard the plate in the designated place for sterilization.

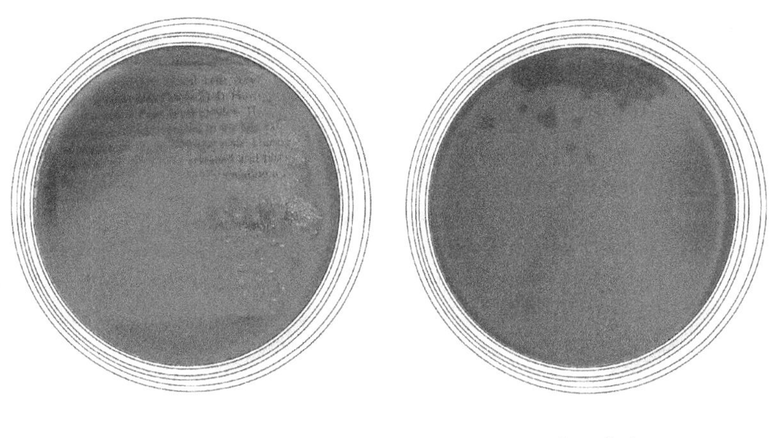

β-hemolysis α-hemolysis.

Appearance of hemolytic reactions on blood agar plates

Troubleshooting

- If possible, incubate in a candle jar; streptococci and enterococci grow better in an elevated CO_2 atmosphere. Hemolytic reactions are also enhanced.

SELECTIVE AND/OR DIFFERENTIAL MEDIA

Eosin Methylene Blue (EMB) Agar

Purpose:

- To isolate and differentiate gram-negative enteric bacilli.

EMB agar contains the dyes eosin and methylene and the sugars lactose and sucrose. EMB agar is a **selective** medium because the dyes will inhibit growth of many gram-positive bacteria.

This medium is used primarily for the preliminary identification of enteric bacilli, that is, bacteria that inhabit and infect the intestinal tract. EMB agar is a **differential** medium that distinguishes between those bacteria that ferment lactose and/or sucrose and those that do not.

If bacteria growing on EMB agar ferment lactose and/or sucrose and produce acid, the lower pH will cause the dyes to precipitate on the colonies. Significant acid results in dark blue colonies with a greenish metallic sheen. Colonies of *Escherichia coli* typically appear like this. Lower acid production from fermentation of the sugars by some bacteria yields pink colonies because less dye is precipitated. Colonies formed by bacteria unable to ferment these disaccharides are colorless or take on the color of the medium.

Materials:

- EMB agar plate
- a specimen suspected of contamination with gram-negative enteric bacilli
  or
  a pure culture of bacteria growing in broth culture.

Procedure:

1. Label the EMB agar plate with the source of the specimen or the name of the bacterium you are testing, your name or initials, and the date.

2. Use aseptic technique to inoculate the EMB agar plate as a streak plate (pp. 43–45). This will isolate individual colonies from a specimen of mixed bacteria.

2a. If you are testing a bacterium for the color of its colonies on EMB agar, use aseptic technique to prepare a streak plate with a loopful of broth culture.

3. Incubate the inoculated plate upside-down for 24–48 hours at 35° C.
4. Examine the plate at 24 and at 48 hours for bacterial growth and the color of the colonies.
5. Discard the plate in the designated place for sterilization.

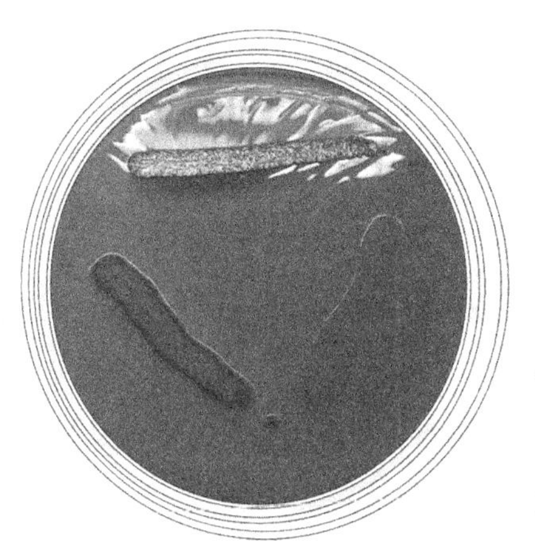

Top - *Escherichia coli*
(dark purple, green metallic sheen)
Lower left - *Enterobacter aerogenes*
(pink)
Lower right - *Salmonella typhimurium*
(colorless)

Appearance of bacterial colonies on EMB agar

Troubleshooting

- EMB plates must be stored and incubated in the dark. Visible light may cause chemical changes in the medium that will interfere with bacterial growth.
- EMB agar is not completely inhibitory for gram-positive bacteria. Staphylococci, streptococci, and enterococci can grow on EMB as very small colorless colonies.

SELECTIVE AND/OR DIFFERENTIAL MEDIA

Mannitol Salt Agar

Purpose:

- To isolate bacteria based on their salt tolerance and differentiate among these isolates for mannitol fermentation.

Mannitol salt agar is **selective** because only bacteria that tolerate 7.5% NaCl will grow on the medium. Among the salt-tolerant gram-positive cocci are species of *Staphylococcus, Micrococcus,* and, to a lesser extent, *Enterococcus*. The addition of the sugar alcohol mannitol makes this medium also **differential**. Acid produced from fermentation of mannitol turns the pH indicator phenol red from red to yellow. Pathogenic staphylococci have this phenotype on mannitol salt agar. Non-pathogenic staphylococci do not have yellow zones around their colonies.

Materials:

- mannitol salt agar plate
- a specimen suspected to be contaminated with staphylococci
  or
  a pure culture of bacteria growing in broth culture
- disposable gloves

Procedure:

1. Label the mannitol salt agar plate with the source of the specimen or the name of the bacterium you are testing, your name or initials, and the date.

2. If you want to isolate individual colonies from a specimen suspected to be contaminated with staphylococci, use aseptic technique to inoculate the mannitol salt agar plate as a streak plate (pp. 43–45).

2a. If you are testing a bacterium for its ability to grow on mannitol salt agar, use aseptic technique to prepare a streak plate with a loopful of broth culture.

3. Incubate the inoculated plate upside-down for 24 hours at 35° C.

4. Examine the plate for bacterial growth and the color of the medium around the bacteria.

5. Discard the plate in the designated place for sterilization.

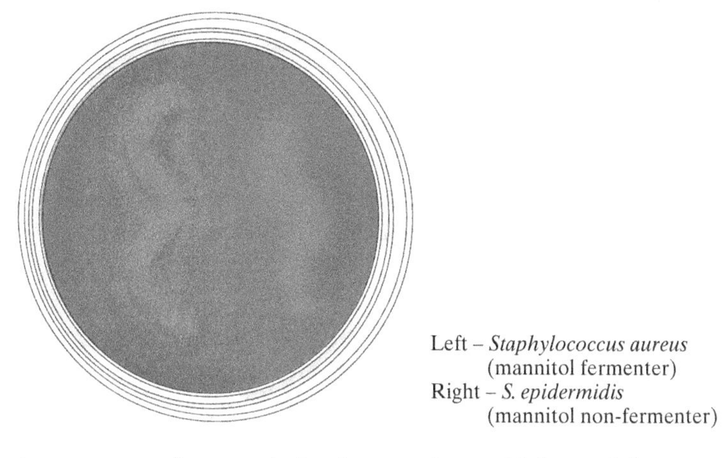

Appearance of a mannitol-salt agar plate with bacterial growth

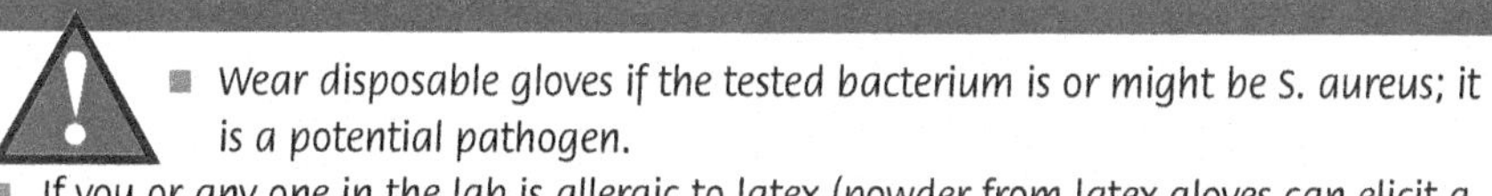

- *Wear disposable gloves if the tested bacterium is or might be S. aureus; it is a potential pathogen.*
- *If you or any one in the lab is allergic to latex (powder from latex gloves can elicit a severe allergic reaction in susceptible people), wear gloves made of nitrile.*

SELECTIVE AND/OR DIFFERENTIAL MEDIA

MacConkey Agar

Purpose:

- To detect and differentiate among gram-negative enteric bacilli, based on their ability to grow on the medium and to ferment lactose.

MacConkey agar plates are often used to process specimens that contain or are contaminated with fecal matter. MacConkey agar is **selective** because gram-positive bacteria are inhibited by crystal violet and bile salts included in the medium but gram-negative bacteria are not affected. Lactose is added to the medium to make it **differential**. Lactose-fermenting bacteria are distinguished from non–lactose fermenters by the color of their colonies. Bacteria that utilize lactose and produce acids will have colonies that are pink to red in color because of a color change in neutral red, the pH indicator that is also in the medium. Colonies of non–lactose-fermenting bacteria are colorless.

Materials:

- MacConkey agar plate
- a specimen with a mixture of gram-positive and gram-negative bacteria
  or
  a specimen containing or contaminated with fecal matter
  or
  a pure culture of a gram-positive bacterium and a pure culture of a gram-negative bacterium in a broth medium

Procedure:

1. Label the MacConkey agar plate with the source of the specimen or the name of the bacterium you are testing, your name or initials, and the date.

2. Use aseptic technique to inoculate the MacConkey agar plate as a streak plate (pp. 43–45). This will isolate individual colonies from a specimen with mixed bacteria or a specimen containing or contaminated with fecal matter.

2a. If you are testing for the ability of gram-positive and gram-negative bacteria to grow on MacConkey agar, use aseptic technique to inoculate the

MacConkey agar plate by streaking a short line on the agar surface with a loopful of broth culture.

3. Incubate the inoculated plate upside-down for 24–48 hours at 35° C.
4. Examine the plate each day for the color of bacterial colonies.
5. Discard the plate in the designated place for sterilization.

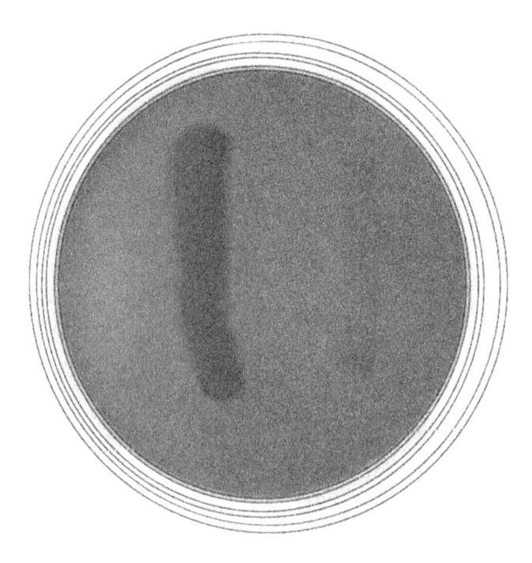

Left – lactose-fermenter
Right – non-lactose fermenter

Appearance of bacterial growth on MacConkey agar

SELECTIVE AND/OR DIFFERENTIAL MEDIA

Phenylethyl Alcohol (PEA) Agar

Purpose:

- To isolate gram-positive bacteria from a specimen containing a mixture of gram-positive and -negative bacteria.

Phenylethyl alcohol (PEA) agar is a **selective** medium. PEA inhibits or slows the growth of gram-negative bacteria by interfering with DNA synthesis. However, gram-positive bacteria will grow in the presence of PEA.

Materials:

- PEA agar plate
- specimen with a mixture of gram-positive and gram-negative bacteria
  or
  a pure culture of a gram-positive bacterium and a pure culture of a gram-negative bacterium in a broth medium

Procedure:

1. Label the PEA agar plate with the source of the specimen or the name of the bacterium you are testing, your name or initials, and the date.

2. Use aseptic technique to inoculate the PEA agar plate as a streak plate (pp. 43–45). This will isolate individual colonies from a specimen with mixed bacteria.

2a. If you are testing for the ability of gram-positive and gram-negative bacteria to grow on PEA agar, use aseptic technique to inoculate the PEA agar plate by streaking a short line on the agar surface with a loopful of broth culture.

3. Incubate the inoculated plate upside-down for 24–48 hours at 35° C.

4. Examine the plate each day for bacterial growth, including relative abundance.

5. Discard the plate in the designated place for sterilization.

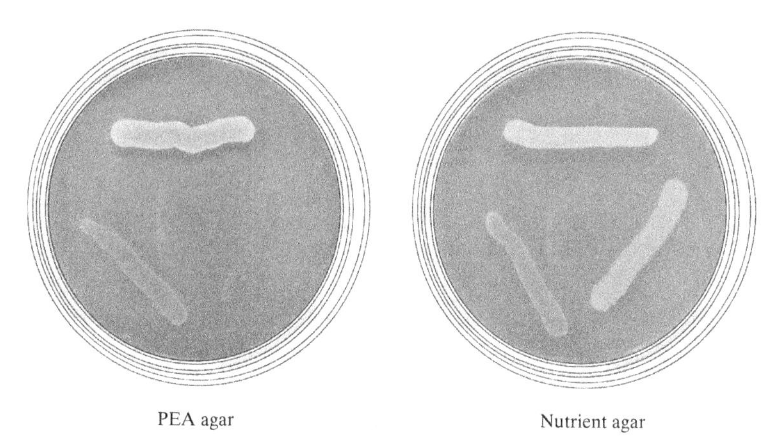

Top - *Staphylococcus aureus* (gram-positive)
Lower left - *Enterococcus faecalis* (gram-positive)
Lower right - *Escherichia coli* (gram-negative)

Appearance of bacterial growth on PEA and nutrient agar; notice the difference in abundance of growth

- *If you prepare PEA agar from dehydrated medium, use caution. PEA is irritating to the eyes, the respiratory system, and the skin. It may cause harm to an unborn child. Do not breathe the dust. Seek immediate medical help if you experience any breathing difficulty from inhaled dust. Wear gloves and eye protection. If you get PEA-containing dust in your eyes or on your skin, immediately flush with lots of water.*

Direct Counting with the Petroff-Hausser Chamber

Purpose:

- To determine the concentration of bacterial cells in a sample by direct counting.

The Petroff-Hausser counting chamber is a thick glass slide with an engraved grid in the middle of its mirrored surface.

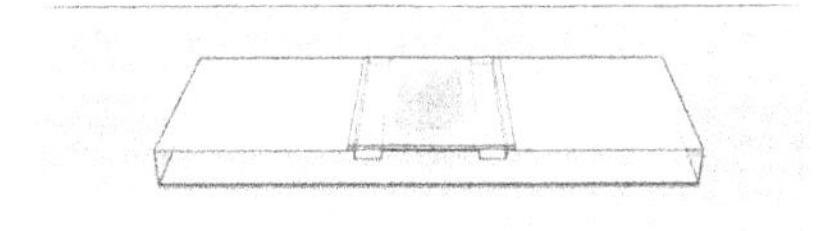

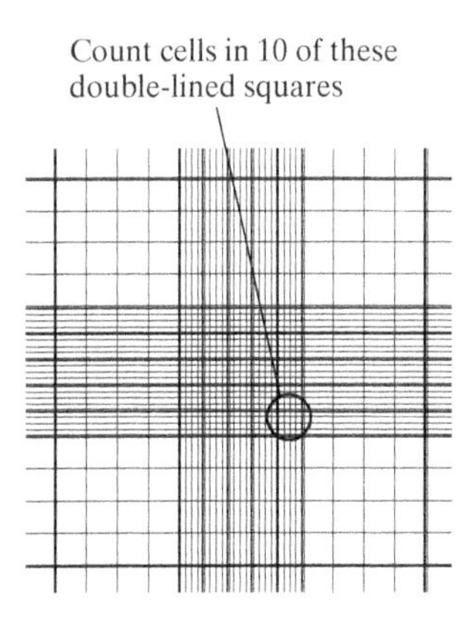

Grid on Petroff-Hausser counting chamber viewed at 40X

When a special coverslip is placed over the grid, it forms a very small chamber of known volume with respect to the squares in the grid.

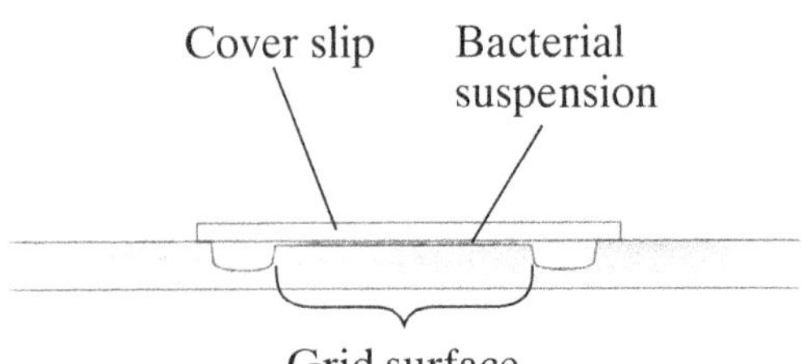

After a bacterial suspension is added under the coverslip, the cells are counted in several of the central double-lined squares.

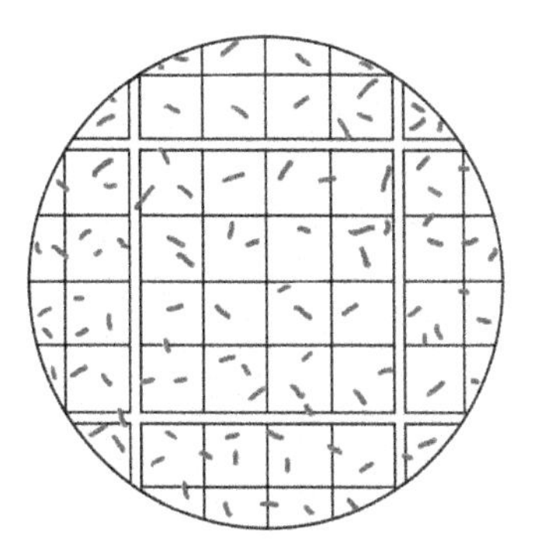

Bacteria viewed at 400X in a
Petroff-Hausser counting chamber

Because the central portion of the grid has 25 double-lined squares, an average cell number is calculated for one square after counting the cells in at least 10 double-lined squares. The concentration of bacterial cells in the hypothetical single square above is calculated as follows:

$$\begin{aligned} \text{cell/mL} &= (\text{average \# of cells in a square})(25\ \text{squares})(50)(10^3) \\ &= 27 \times 25 \times 50 \times 10^3 \\ &= 3.4 \times 10^7 \end{aligned}$$

The value 50 is 1/height of chamber (0.02 mm). The value 10^3 is the conversion factor $1/\text{mm}^3$.

A drawback to direct counting is the inability to discriminate between live and dead cells; you can only determine the total cell concentration. Fluorescent dyes are available for distinguishing live and dead cells, but a fluorescence microscope is required.

Materials:

- Petroff-Hausser counting chamber (slide and special coverslip)
- Pasteur pipette, sterile, with bulb
- microscope
- hand-held cell counter
- 10% bleach solution or 70% ethanol
- bacterial suspension
- tubes with 9 mL 3.7% formaldehyde in 0.85% NaCl (if needed)
- pipettes, 1 mL, sterile (if needed)

Procedure:

1. If the bacteria are motile, prepare a 1/2 dilution in 3.7% formaldehyde/0.85% NaCl before you begin.

2. Using capillary action from a Pasteur pipette, add a small volume of bacterial suspension under the coverslip on the Petroff-Hausser slide.

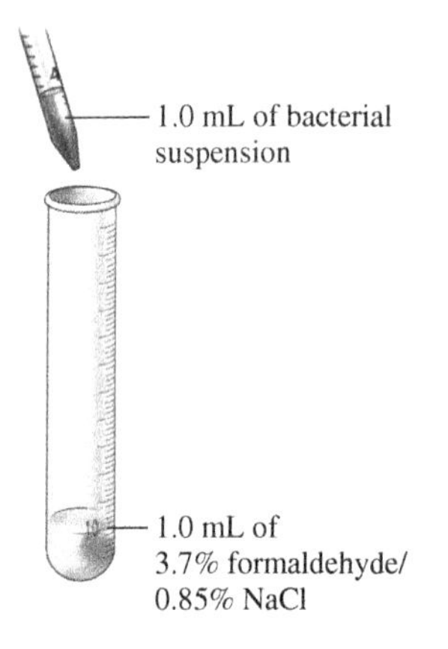

3. Allow the counting chamber to stand for a few minutes to allow the cells to settle, then gently place it on the microscope stage. With the 4X lens in place, center the grid over the light.

4. Switch to the 40X lens. Center one of the double-lined squares in the field of vision. Count the cells in one single-lined square, including those touching a line at the top and left side. Do not count cells touching a line at the bottom or right side. Continue counting the cells in ten squares.

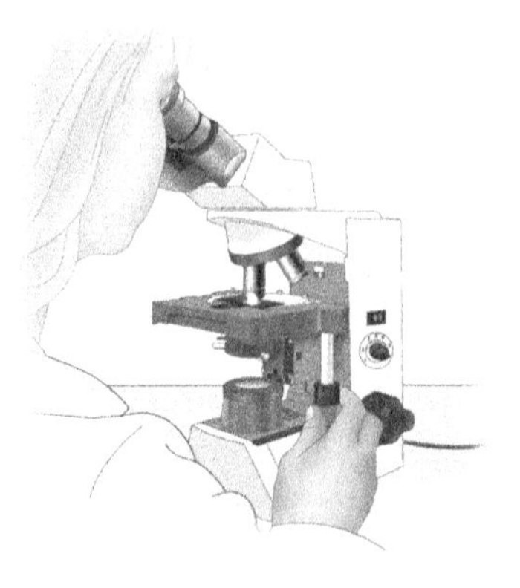

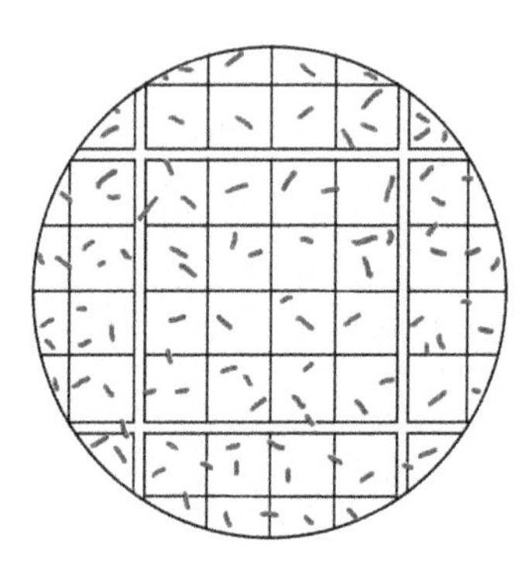

5. If the cell number is too large for accurate counting, prepare a 1/10 dilution as described in "Preparing a Standard Plate Count" (pp. 183–185). Then proceed as described in Steps 1–3.

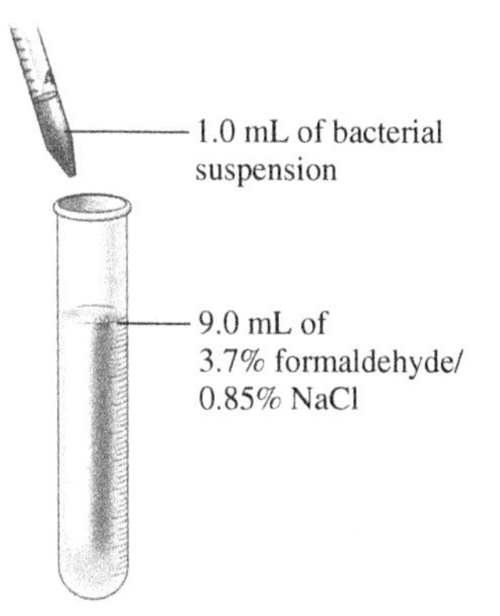

6. Calculate cells/mL of the original bacterial suspension, as described on page 179. If you prepared a dilution, also multiply by the inverse of the dilution, e.g., for a 1/2 dilution, multiply by 2.
7. If the bacteria are viable, disinfect the slide and coverslip in 10% bleach or 70% ethanol. After soaking for 15 minutes, rinse with distilled water. Use lens paper to dry.

- *If the bacteria are alive, use aseptic technique to load the counting chamber. When you have completed counting a sample, disinfect the coverslip and slide.*

Troubleshooting

- Bacteria are best viewed if the light intensity is minimal, and you can miss cells if the light intensity is too bright. Adjust the light level with the diaphragm lever to a low intensity that provides a contrast between the cells and the background.
- A No. 2 glass coverslip, the type commonly used, will sag when placed over the grid. This will decrease the volume in the counting chamber and thus reduce the cell count per square.
- If you add too great a volume of cell suspension, it will overflow and create an inaccurate volume under the coverslip. This can be avoided by adding the fluid slowly.

- If you count the cells in only a few squares, you will not obtain an adequate sampling of cell population size.
- The cell concentration in the sample must be large enough ($\geq 10^7$/mL) for accuracy of counting.
- Cells that are non-motile will look like they are "shivering." This is Brownian movement, caused by rapidly moving water molecules continually colliding with the cells.
- Rinse the coverslip and slide with 75% ethanol after you have completed your work, even if the bacteria have been killed.
- Use lint-free lens paper to wipe the coverslip and slide. Kimwipes and paper towels will scratch the special surface around the grid.

PREPARING A STANDARD PLATE COUNT OF VIABLE BACTERIA

Purpose:

- To determine the number of viable bacteria in a culture or in food, water, or soil.

The procedure preparing a standard plate count of viable bacteria in a sample comprises three stages: 1) prepare a serial dilution of the sample; 2) mix an aliquot of dilution with melted agar that is then poured in a plate, cooled, and incubated; 3) count the number of colonies embedded within and on the agar, and use the number(s) from the plate(s) with >30 and <300 colonies to calculate the **colony-forming units (CFU)** present in the sample.

Materials:

- tubes with 9.0 mL sterile 0.85% NaCl
- pipettes, sterile, 1.0 mL
- pipettor
- tubes containing melted plate count agar, 18 mL, kept in a water bath at 48–50° C *or*
    - tubes containing plate count agar, 18 mL
    - beaker, 600 mL, half-filled with distilled water
    - thermometer
    - hot plate
- Petri plates, sterile
- sample containing bacteria

An inexpensive pipettor can be easily constructed for use with disposable 1 mL pipettes.

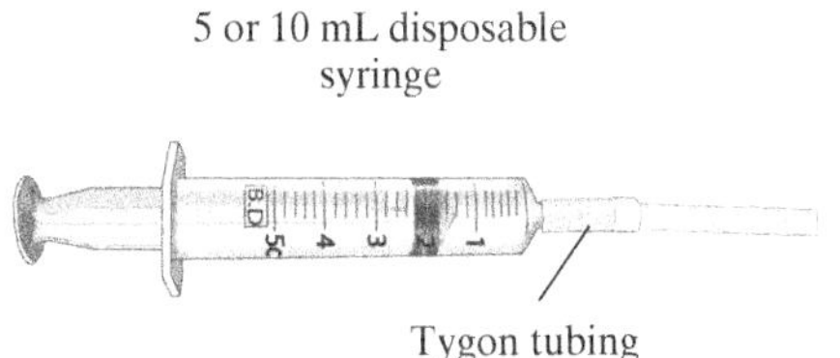

Procedure:

1. If the agar is not already melted, heat the tubes in a beaker with water on a hot plate until the agar is liquefied. Cool the tubes to about 48–50° C.

2. Label 7 saline tubes 10^{-1} through 10^{-7}; these are **dilution tubes**. Label the bottoms of 5 empty Petri plates 10^{-4} through 10^{-8}. Also include your name or initials and the date.

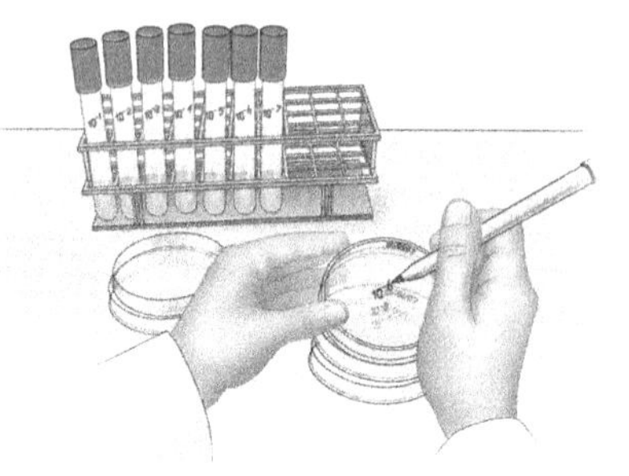

3. If the sterile 1.0 mL pipettes are in a canister, remove the lid and gently shake the canister until one pipette extends further than the others. Remove this pipette without touching any of the others.

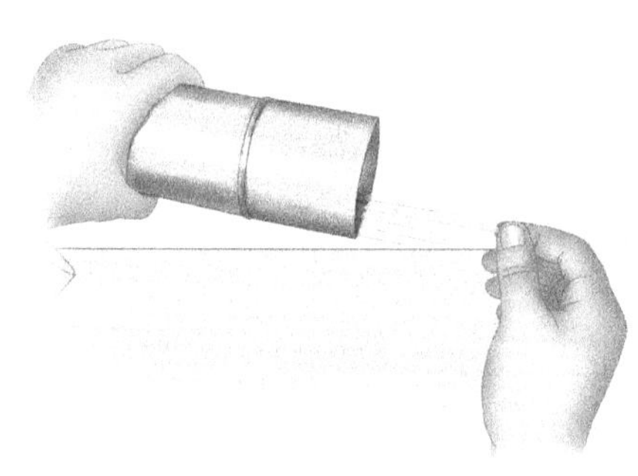

4. If the sterile 1.0 mL pipettes are individually wrapped, open the end of the wrapper where the pipette inside has a cotton plug.

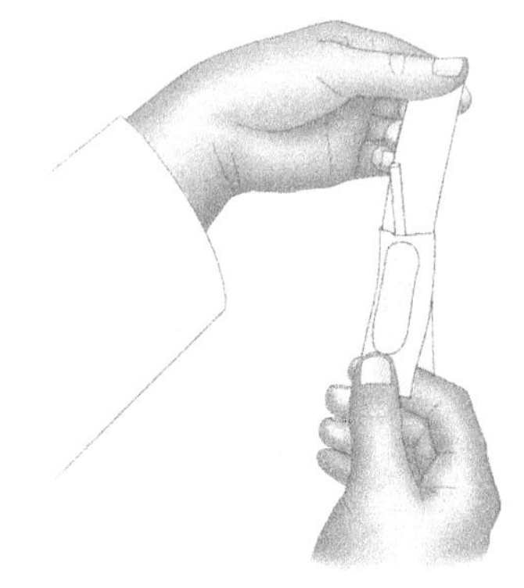

5. Use aseptic technique to transfer 1.0 mL of sample to the 10^{-1} dilution tube.

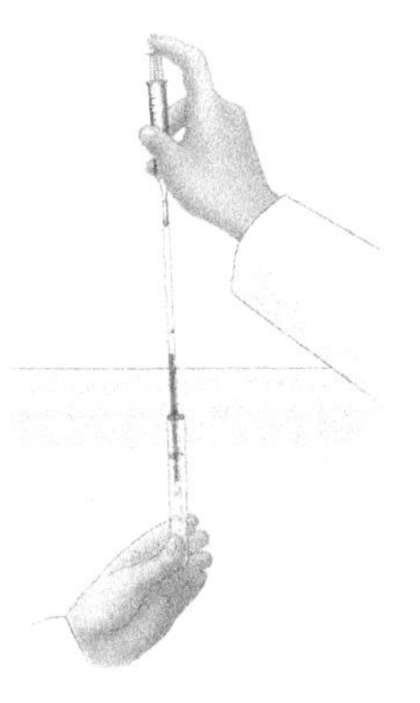

6. Place the pipette in the proper container for sterilization.

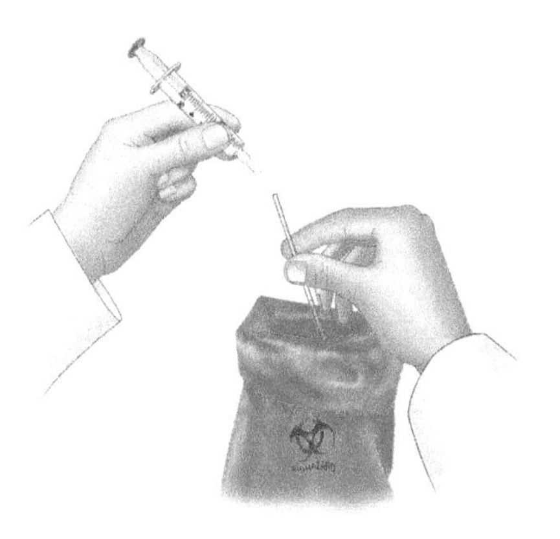

7. Mix the contents of the tube by vigorously rolling it between your open palms about 25 times.

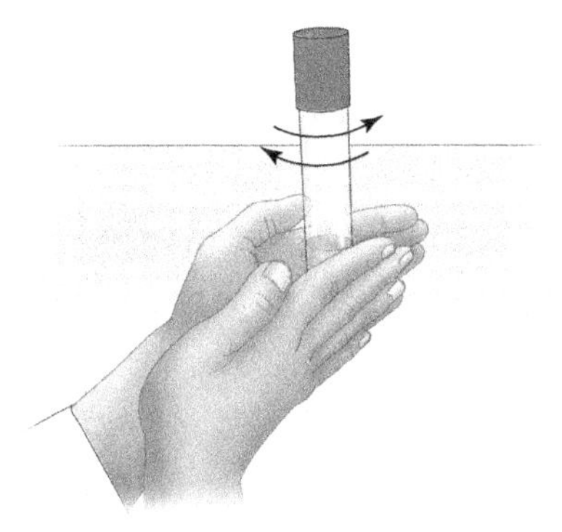

8. Repeat this process, transferring 1.0 mL from tube 10^{-1} with a fresh pipette into tube 10^{-2}. Follow Steps 2–6 until 10^{-8} is prepared. Use a fresh pipette for each transfer.

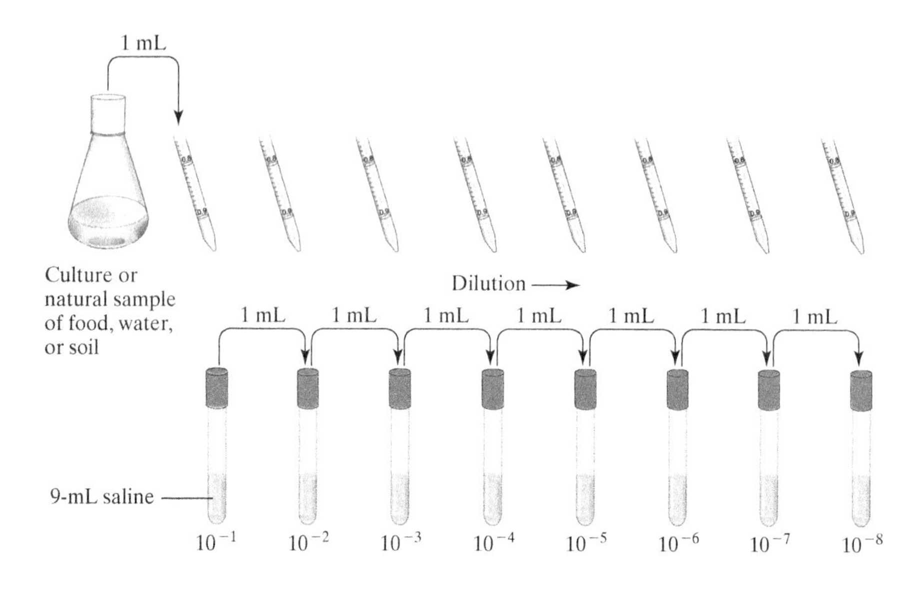

9. Use aseptic technique to transfer 1.0 mL from tube 10^{-8} to a tube with melted agar. Save the pipette, taking care not to touch it to anything.

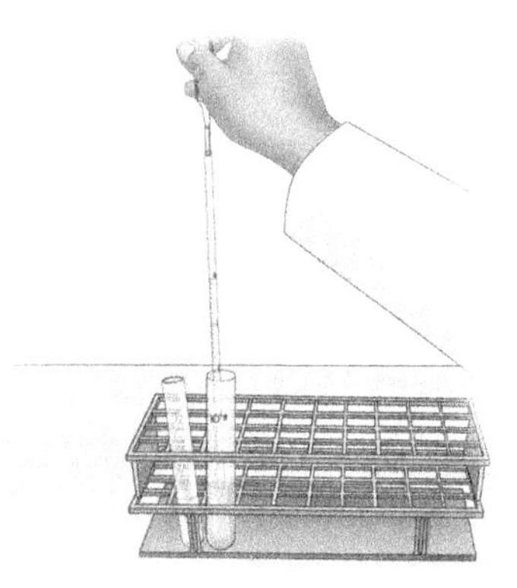

10. Mix the contents of the tube by vigorously rolling it between your open palms about 25 times.

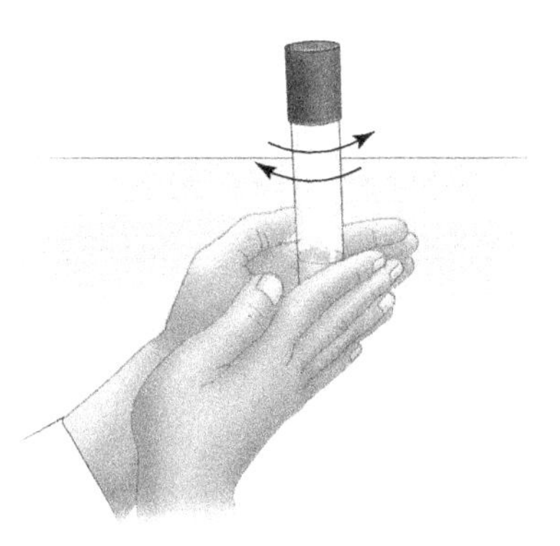

11. Use aseptic technique to pour the contents of the agar tube into the plate marked 10^{-8}.

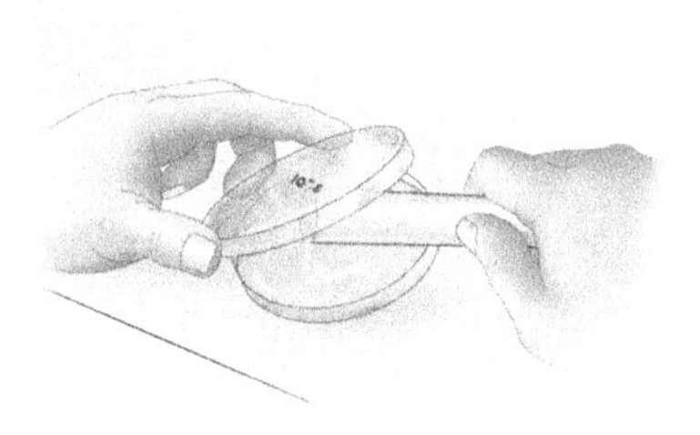

12. Spread the melted agar over the bottom of the plate by gentle swirling. Avoid splashing over the edge or onto the lid.

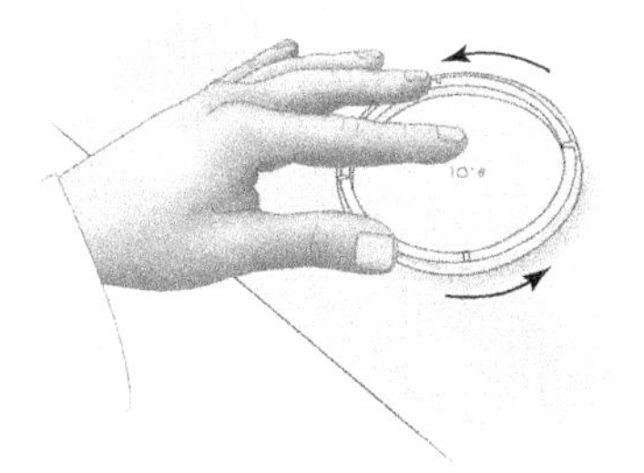

13. Use the pipette from Step 9, and, in order of increasing concentration (descending tube/plate number), transfer 1.0 mL from each dilution tube into a separate tube with melted agar, as described in Steps 10–12.

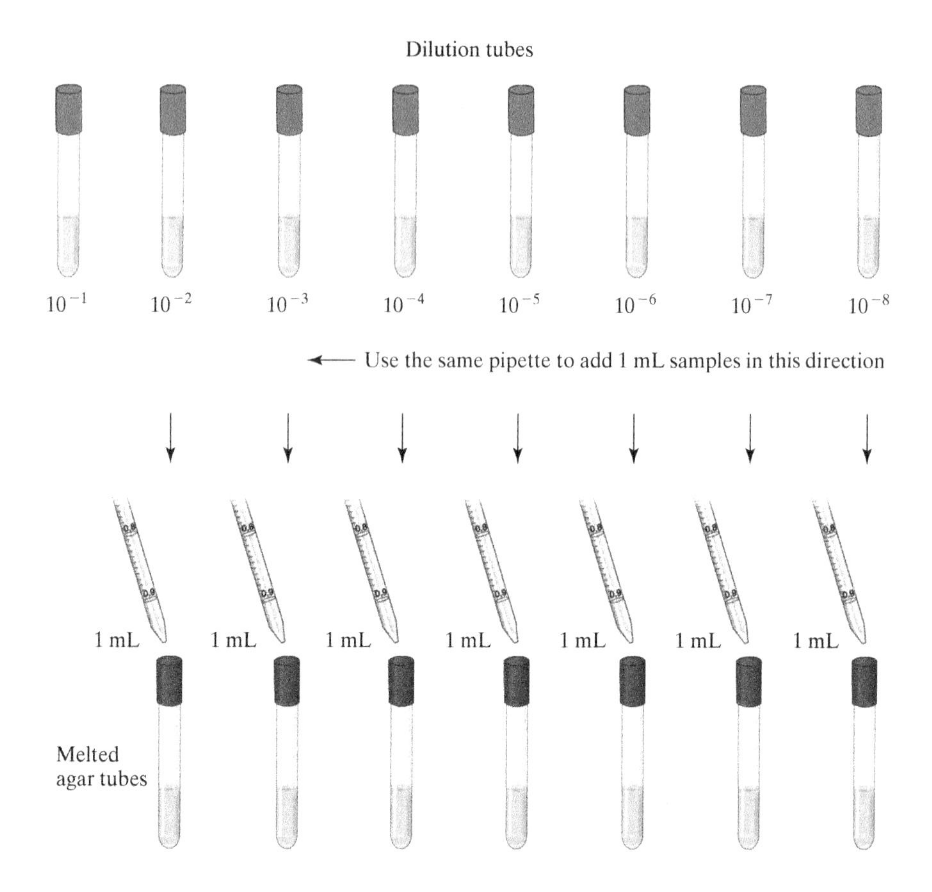

14. After the agar in the plates has solidified, incubate them upside-down at 35° C for at least 24 hours.

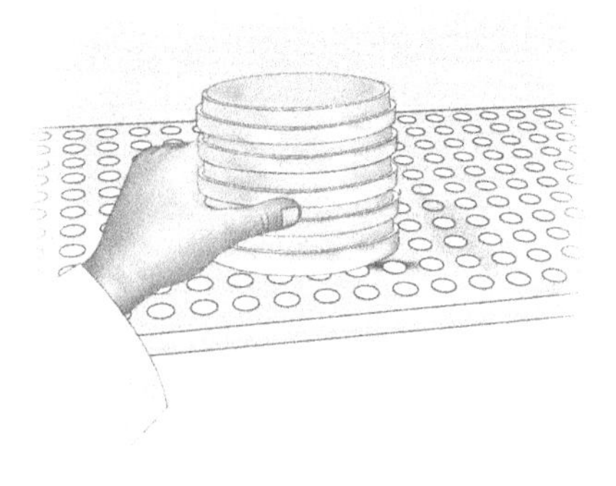

15. View the plates with the bottom side up. Count the number of colonies (on the surface and embedded within the agar) of any plate that has 30–300 colonies. Touch the plate bottom with a fine point marker as a reminder that you have counted that colony.

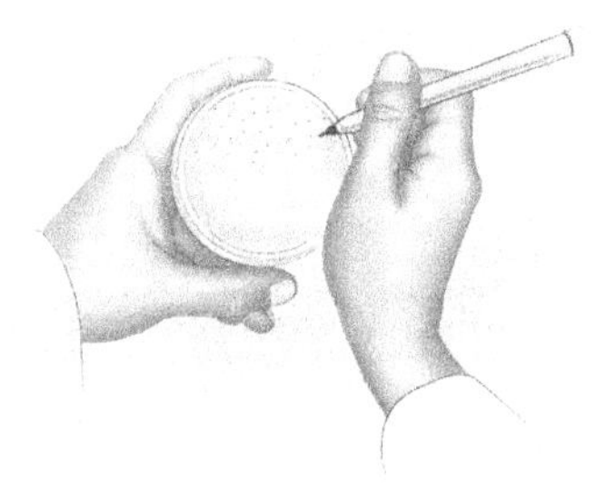

16. Discard the plates in the designated place for sterilization.

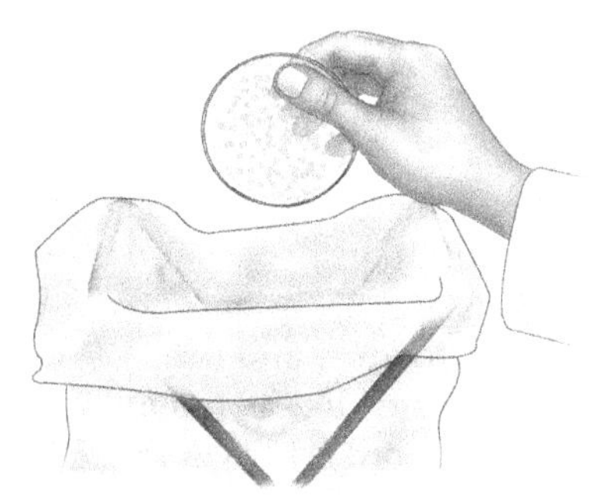

To calculate the concentration of viable cells in the sample (CFU/mL), use the following equation:

$$\frac{\text{CFU}}{\text{mL}} = \frac{\text{number of colonies on plate}}{\text{dilution of sample that was plated}}$$

For example, suppose that 187 colonies were counted on the 10^{-5} plate. When these numbers are used in the equation,

$$\frac{\text{CFU}}{\text{mL}} = \frac{187}{10^{-5}} = 187 \times 10^{5}$$

$$= 1.9 \times 10^{7}$$

Because the plate count method has at least a 10% margin of error, only two significant numbers appear in the final value.

- *Never use your mouth to draw the bacterial suspension or any dilution into a pipette or tube: use a mechanical pipetting device.*
- *Do not set a contaminated pipette down on the lab bench top. When you have finished using it, place the pipette in the proper place for sterilization.*
- *Place tubes that have been emptied of agar in the designated place for sterilization.*

Troubleshooting

- Use the bottom of the meniscus when you measure the volume of a solution with a pipette.
- Keep the pipette vertical to see the meniscus accurately.
- Know the type of pipette that you are using. Check the tip to distinguish between a serological and a measuring pipette. The volume delivered by a serological pipette is the difference between the location of the meniscus on the volume markings and the very tip of the pipette. A measuring pipette delivers the difference between a zero marking near the tip and the meniscus.
- Be consistent in how you measure volumes for each dilution and how you mix each dilution.
- If you're using the same pipette to transfer bacteria, remember to work from lower to higher concentrations.
- Only bacteria that can withstand the brief exposure to the temperature of the melted agar will survive to grow and form colonies.
- Allow the agar tubes that were melted in boiling water to cool, but make sure the agar remains liquid in order to avoid premature gelling and unwanted bacterial death. It is best to keep melted agar tubes in a hot water bath until you need to use the liquefied agar.
- Work quickly so that the melted agar in the tube does not solidify before you pour the mixture into a sterile Petri plate.
- Avoid vigorous swirling of the melted agar to avoid splashing over the edge or onto the lid of the plate.
- Pour the melted agar carefully but quickly into the plate to prevent bubbles.
- Agar will become translucent and lighter in color when it has solidified.
- The term "colony-forming unit" refers to either a single cell or a clump of cells that serve as the founder of a colony.
- Check the far edges of the agar for colonies that are easy to miss.
- In a research project, triplicate plates are prepared for each dilution. To calculate CFU/mL, use the average number of colonies for that dilution in the equation.

USING TURBIDIMETRY TO ESTIMATE CELL DENSITY

Purpose:

- To estimate the cell density in a suspension of bacteria by correlating cell concentration to light absorbance, as measured with a spectrophotometer.

When the population density of bacteria growing in broth reaches about 10^7 cells/mL, the broth appears turbid. Less light comes out than enters because light passing through the suspension is now scattered by the more crowded cells. A spectrophotometer measures the lowered transmittance of light or the increased **absorbance**.

Absorbance is directly proportional to cell density. Correlation of absorbance with known concentrations of bacterial cells is experimentally measured to produce a standard curve. Live and dead cells are not distinguished in this technique.

Materials:

- spectrophotometer
- cuvette tubes for spectrophotometer, clean
- pipettes, 5 mL, sterile
- pipettor
- Kimwipes
- broth growth medium, 20 mL, sterile
- bacterial culture in growth broth
- disinfectant (10% bleach) in a beaker
- Pasteur pipettes, sterile, with bulbs
- test tubes, sterile
- materials for either "Direct Counting with the Petroff-Hausser Chamber" or "Preparing a Standard Plate Count"

Procedure:

Preparing a two-fold dilution of a bacterial suspension

1. Use aseptic technique to add 4 mL of sterile broth to each of 4 sterile tubes labeled 1/2, 1/4, 1/8, and 1/16.

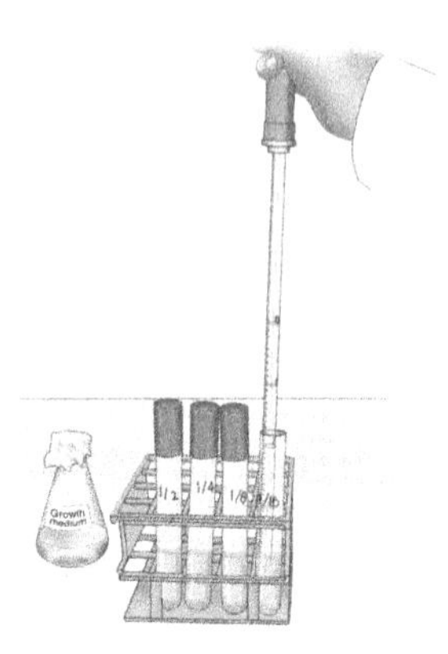

2. Add 4 mL of the bacterial suspension to the tube marked 1/2. Discard this pipette in the designated place for sterilization. Take a fresh pipette and draw the contents up and down four times to mix.

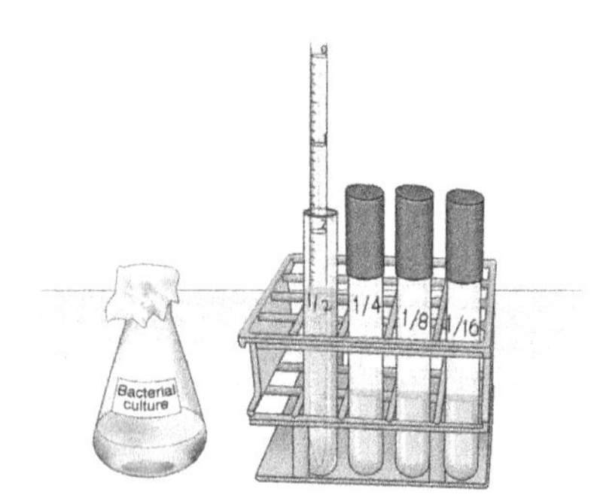

3. Continue the two-fold serial dilution through tube 1/16 as described in Steps 1 and 2.

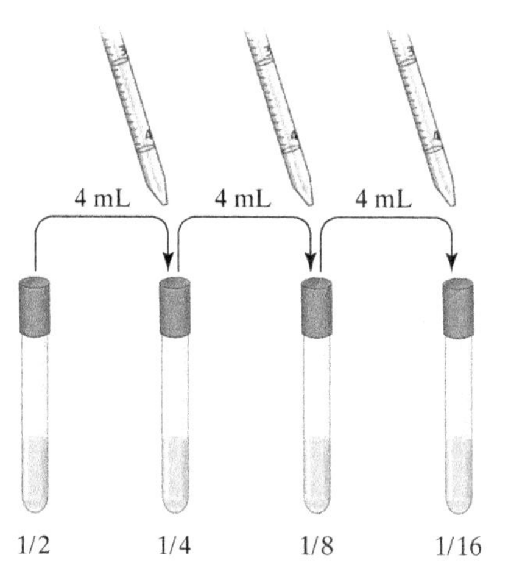

Calibrating the spectrophotometer (Spectronic 20)

4. Plug in the spectrophotometer. Turn it on with the power/zero control knob.

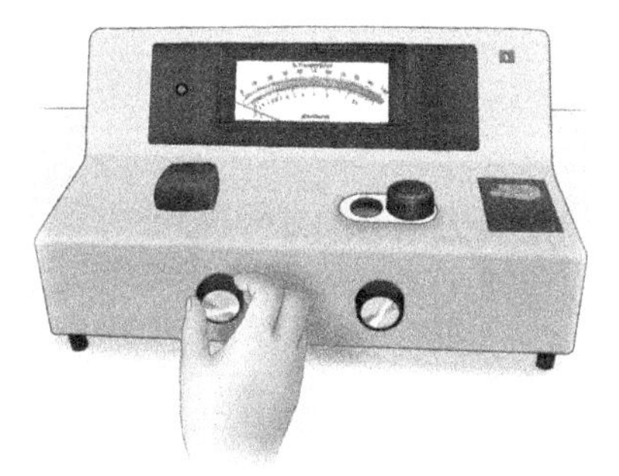

5. Set the wavelength control to 600 nm. Allow the unit to warm up for 10 minutes.

6. Turn the power/zero control knob (to the left) until the transmittance reads 0%. Make sure that the cuvette holder is empty and the lid is closed.

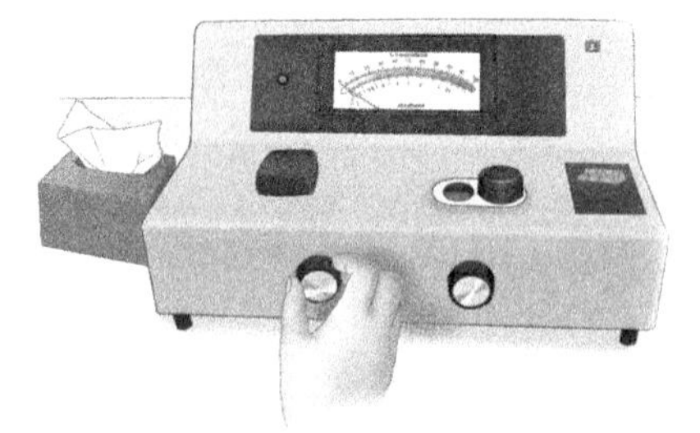

7. Add 4 mL of sterile growth medium to a clean cuvette, using a sterile 5 mL pipette. This is the **blank cuvette**.

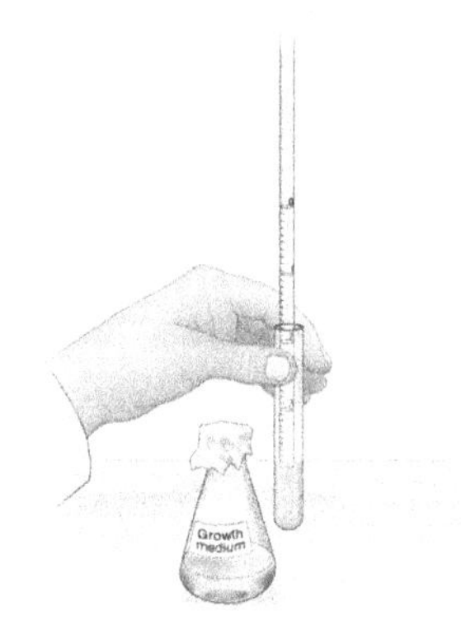

8. Wipe the outside of the blank cuvette with a Kimwipe. Insert the cuvette all the way into the cuvette holder. Align the reference line on the cuvette to the mark at the front of the holder. Close the cover over the holder.

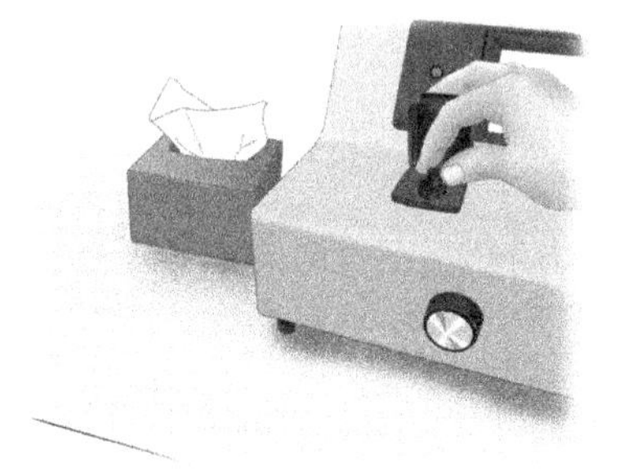

9. Turn the light control knob (right) until the transmittance is 100% (absorbance is 0).

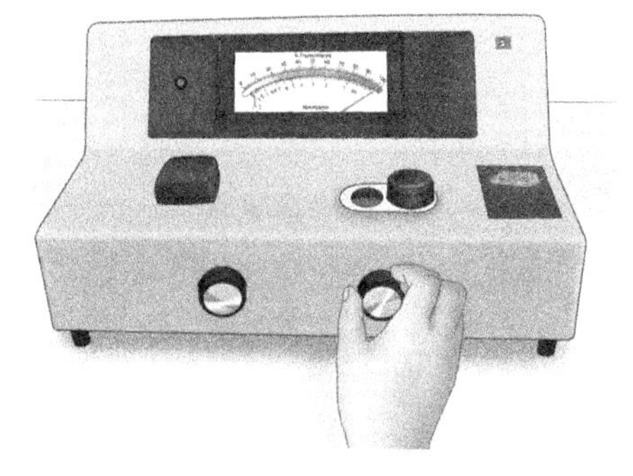

10. Repeat Steps 6–9 to ensure that the instrument is properly adjusted. Remove and save the blank cuvette for recalibration.

Measuring the absorbance of each dilution

11. Use aseptic technique to add 4 mL of 1/16 dilution of bacterial suspension to a second cuvette. Place the pipette back in the dilution tube; you will use the pipette again.

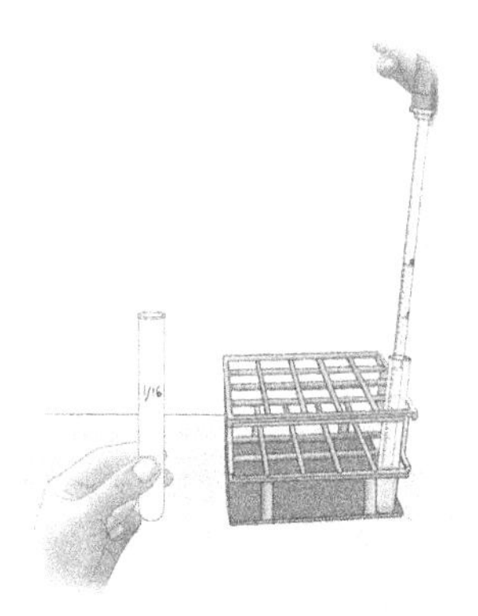

12. Wipe the outside of the cuvette with a Kimwipe. Insert the cuvette into the holder. Align the reference line on the cuvette to the mark at the front of the holder as in Step 8. Close the cover. Read and record the absorbance.

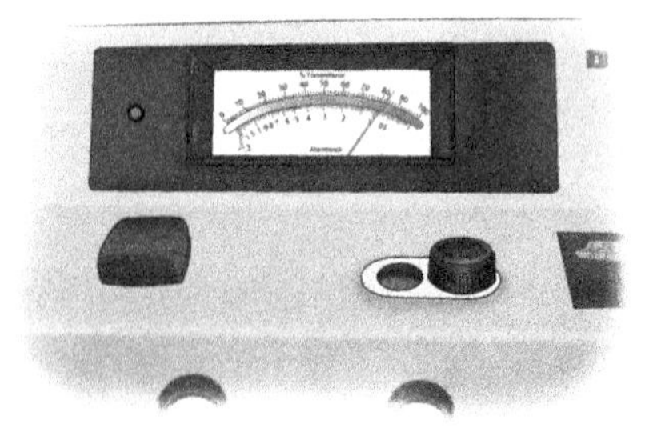

13. After you take the cuvette out of the holder, use a Pasteur pipette and bulb to remove the bacterial suspension. Discard the bacteria in disinfectant.

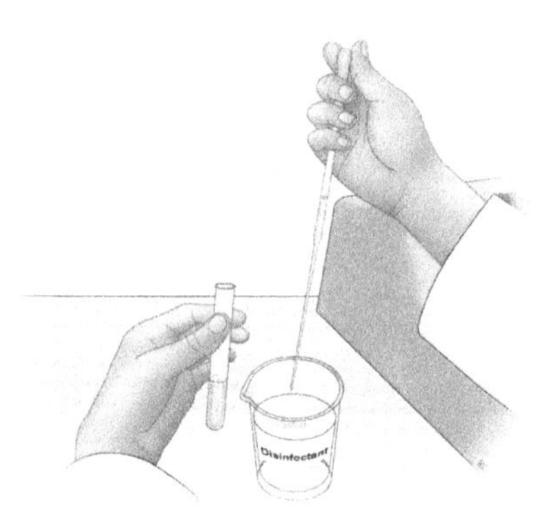

14. Reset the spectrophotometer, following Steps 6–9.

15. Using the same pipette each time, follow Steps 11–14 for dilutions 1/8, 1/4, 1/2, and undiluted. Use the same cuvette for each measurement.

16. Determine the undiluted cell concentration by performing a direct cell count with either the Petroff-Hausser counting chamber (pp. 178–182) or a standard plate count (pp. 183–190).

17. When you have determined the cell concentration in the original broth culture, calculate the cell concentration for each of the dilutions (1/2, 1/4, etc.).
18. Prepare a standard curve in which you plot cell concentration on the x-axis and corresponding absorbance on the y-axis.

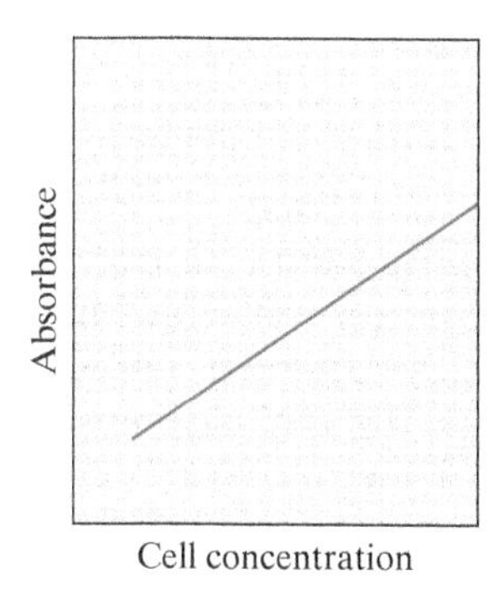

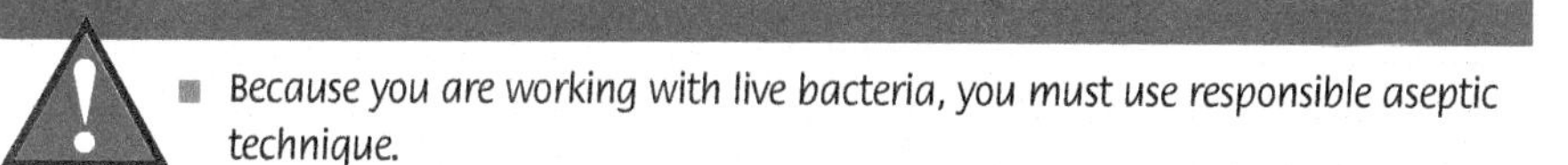

- *Because you are working with live bacteria, you must use responsible aseptic technique.*

Troubleshooting

- Remove fingerprints and moisture from the outside of a cuvette by wiping with a Kimwipe or other lint-free paper just before you insert the cuvette in the holder. Do not use a paper towel because it may scratch the glass.
- If you are using a digital spectrophotometer, be sure to set for reading absorbance with the mode select button.
- If you are using an analog spectrophotometer, read the absorbance value when the needle is superimposed over its image in the mirror in the background.
- Recalibrate the spectrophotometer after each reading to avoid drift.
- Close the cover on the cuvette holder when you take a reading. Extraneous light that leaks in will cause spurious results.
- Be sure to shake the original culture tube so the bacteria are uniformly dispersed.
- Disperse the bacteria in a dilution tube by gently shaking the tube before you add the suspension to the cuvette.
- Read the absorbance immediately after you insert the cuvette, otherwise the bacteria may begin to settle out.
- Remember that this technique provides only an estimate of the total cell count, including live and dead cells, in a sample.
- Use a fresh pipette for preparing each dilution.
- If cost or availability is critical, you can use the same pipette to fill the cuvette when you measure the absorbance of each dilution. But you must begin with the highest dilution, i.e., 1/16.

Determining Bacteriophage Titer by Plaque Assay

Purpose:

- To determine the concentration of infective bacteriophage in a sample by plaque assay.

In a **plaque assay**, bacteriophage and *Escherichia coli* host cells are mixed in a thin layer of low-concentration, or soft, agar. Phage replication and bacterial growth occur together. Progeny bacterial cells will become uniformly distributed throughout the soft agar if growth is undisturbed. This uniform growth is called a bacterial "lawn." However, 30 minutes after a phage infects an *E. coli* cell, the host cell bursts with the release of nearly 100 newly-assembled phage. Many of these new phage will infect nearby *E. coli* cells. After multiple rounds of phage replication, a "hole," called a **plaque**, appears in the bacterial lawn. Each plaque designates the outcome of the original infection of a bacterial cell with a single phage. The number of plaques provides an approximation of the number of virions added to the mixture of soft agar, bacteriophage, and *E. coli* cells. Concentration, or **titer**, of phage is measured in **plaque-forming units (PFUs)** per millimeter.

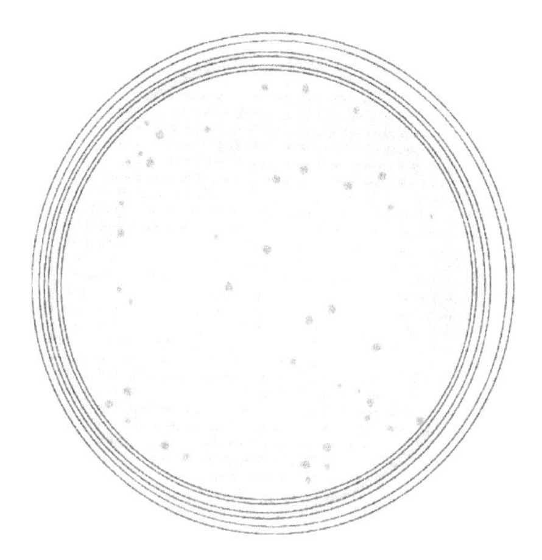

Bacteriophage plaques in a lawn of *Escherichia coli*

Materials:

- 10 tubes with 9.0 mL trypticase soy (TS) broth (dilution tubes)
- 10 Petri plates with nutrient agar (base layer agar plates); plates should be kept warm in the incubator until just before use
- 11 tubes (13 x 100 mm) with 2.0 mL "soft" TS agar (0.7 g agar/100 mL trypticase soy broth), melted and kept in a water bath at 48° C
- *Escherichia coli* broth culture, 4–6 hours old; must be a permissive host strain for the bacteriophage strain
- bacteriophage suspension
- pipettes, sterile, 1.0 mL
- pipettor (see p. 183)
- water bath, 48° C

Procedure:

1. Label the bottoms of 10 agar Petri plates with the appropriate dilution (10^{-1} to 10^{-10}). Include your name or initials and the date. Label 10 dilution tubes 10^{-1} to 10^{-10}.

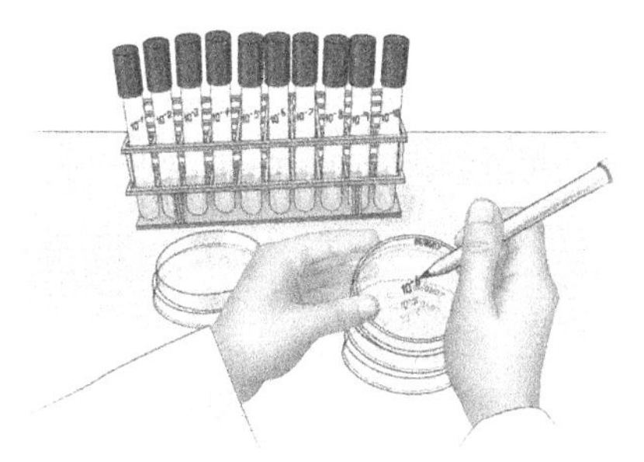

2. If the sterile 1.0 mL pipettes are in a canister, remove the lid and gently shake the canister until one pipette extends further than the others. Remove this pipette without touching any of the others.

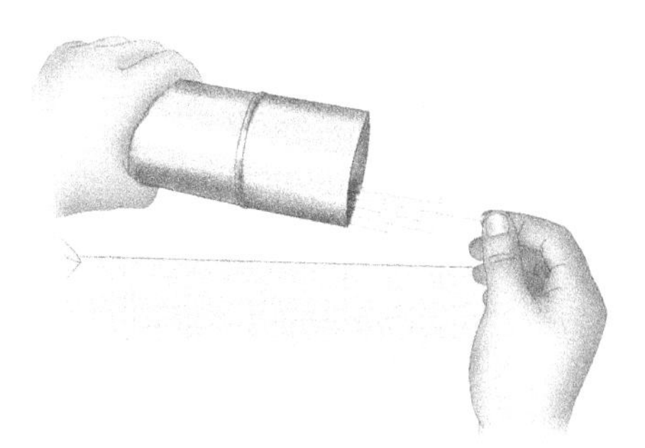

3. If the sterile 1.0 mL pipettes are individually wrapped, open the end of the wrapper where the pipette inside has a cotton plug.

4. Use aseptic technique to transfer 1.0 mL of sample to the 10^{-1} dilution tube.

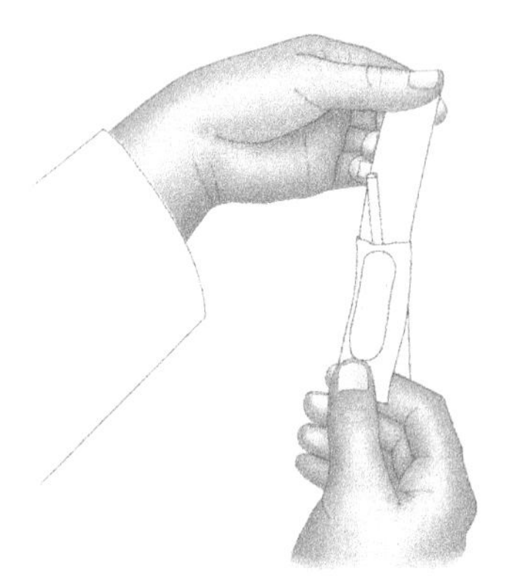

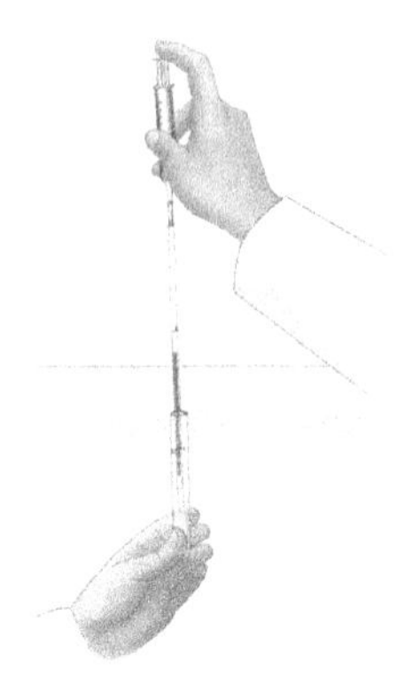

5. Place the pipette in the proper container for sterilization.

6. Mix the contents of the tube by vigorously rolling it between your open palms about 25 times.

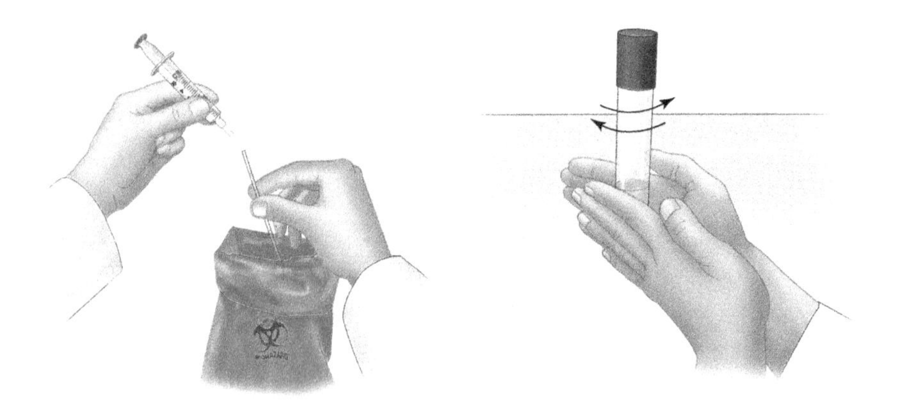

7. Repeat this process, transferring 1.0 mL from tube 10^{-1} with a fresh pipette into tube 10^{-2}. Follow Steps 4–6 until dilution tube 10^{-10} is prepared. Use a fresh pipette for each transfer.

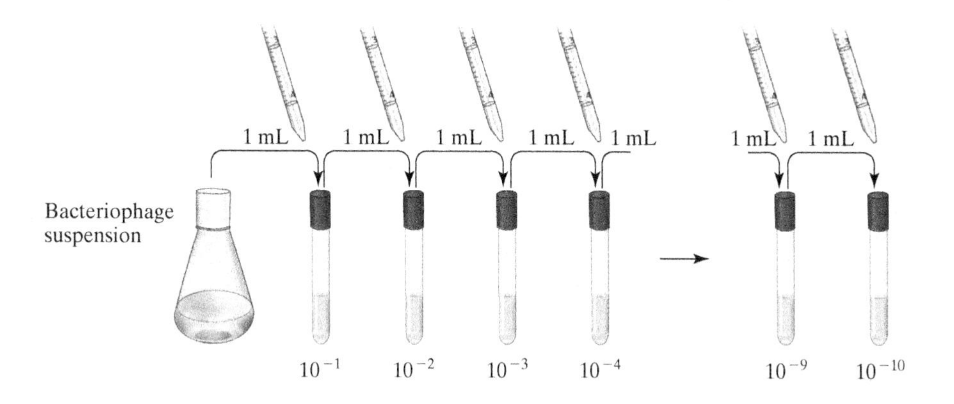

8. Use aseptic technique to add 0.1 mL of the *E. coli* culture to each of the soft agar tubes in the water bath.

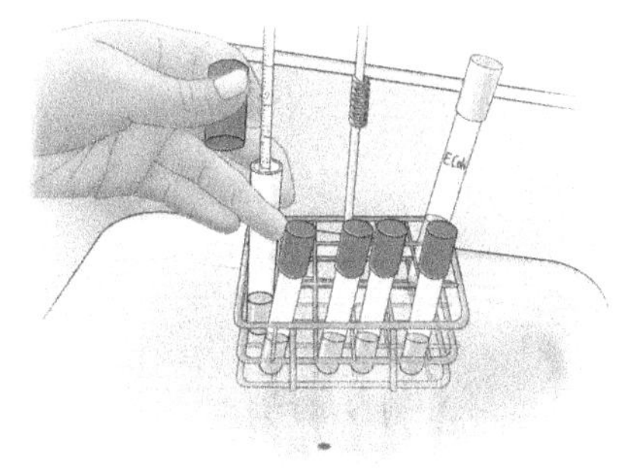

9. Use aseptic technique to transfer 1.0 mL of 10^{-10} dilution tube into a soft agar tube. Save the pipette for additional transfers.

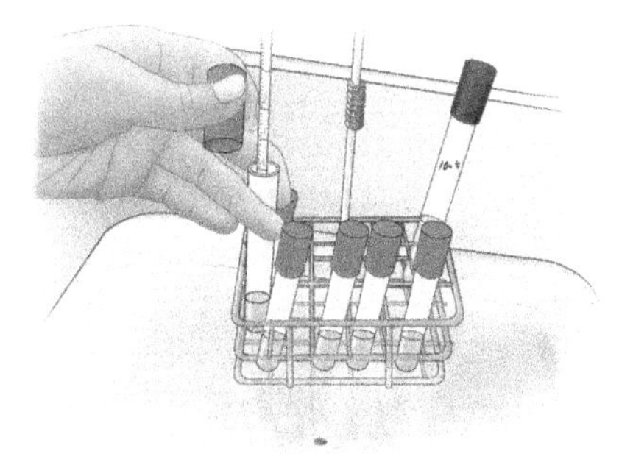

10. Remove the soft agar tube from the water bath. Gently mix the contents to avoid foaming. Quickly pour the contents evenly over the surface of plate 10^{-10}, shaking the tube to empty it completely.

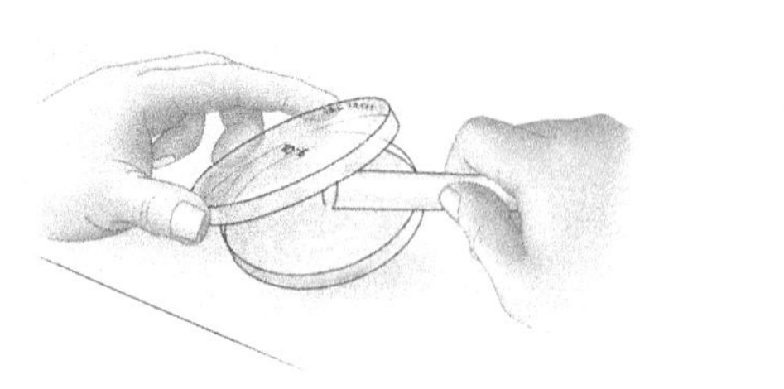

11. Spread the melted agar over the bottom of the plate by gentle swirling. Avoid splashing over the edge or onto the lid. Set the plate aside until the top soft agar layer has solidified.

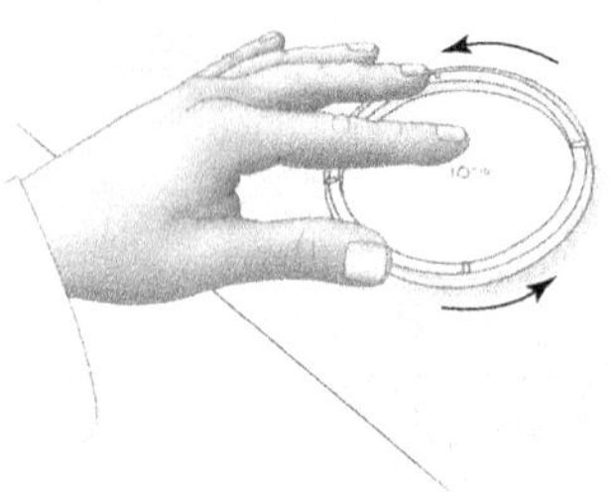

12. Use the pipette from Step 9, and in order of increasing concentration (descending tube/plate number), transfer 1.0 mL from each dilution tube into a separate soft agar tube with *E. coli*, as described in Steps 9–11. Do this through dilution 10^{-4}.

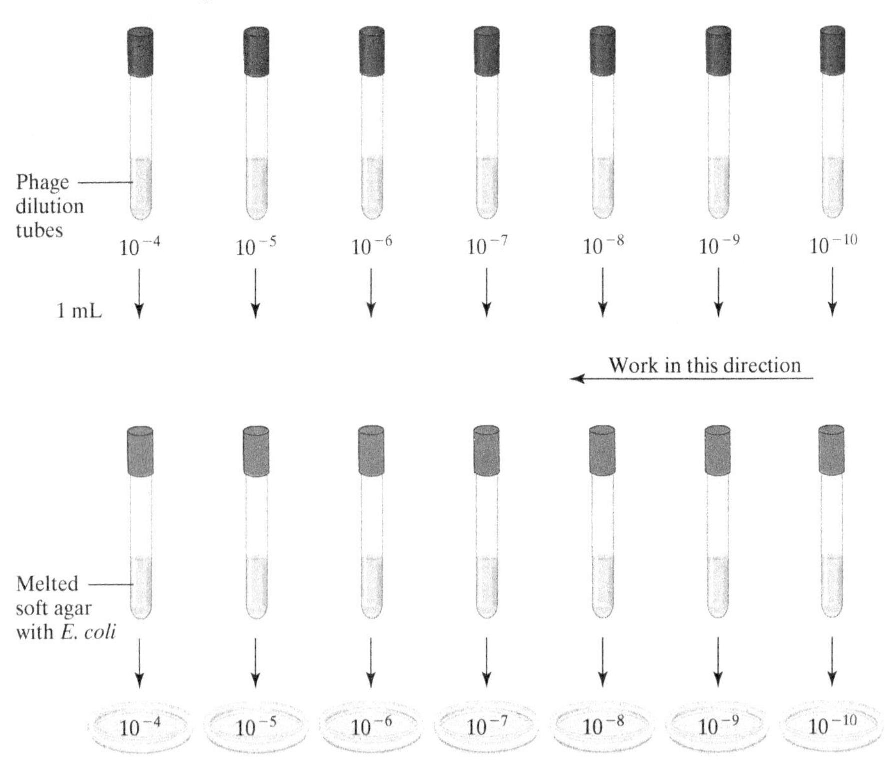

13. After all of the top agar layers have solidified, incubate the plates upside-down at 35° C for 6–8 hours.

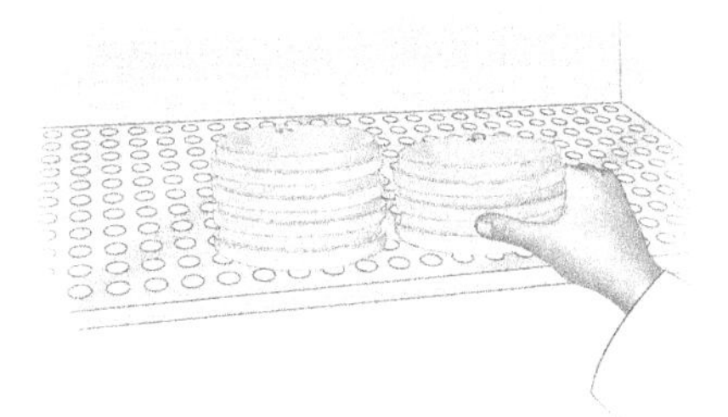

14. View the plates with the bottom side up. Count the number of plaques of any plate that has between 30–300 plaques. Touch the plate bottom with a fine point marker as a reminder that you have counted that plaque.

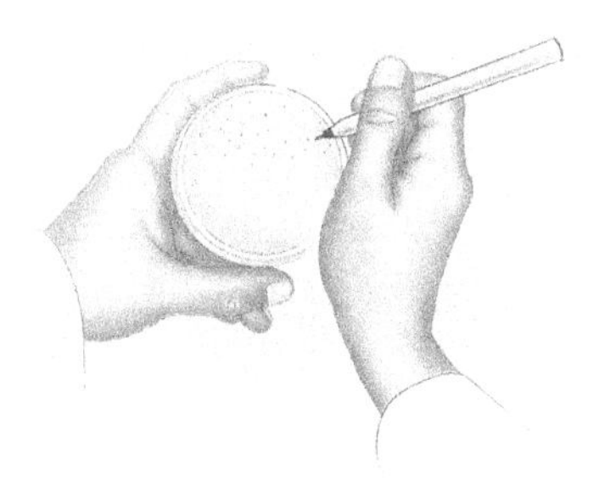

15. Discard the plates in the designated place for sterilization.

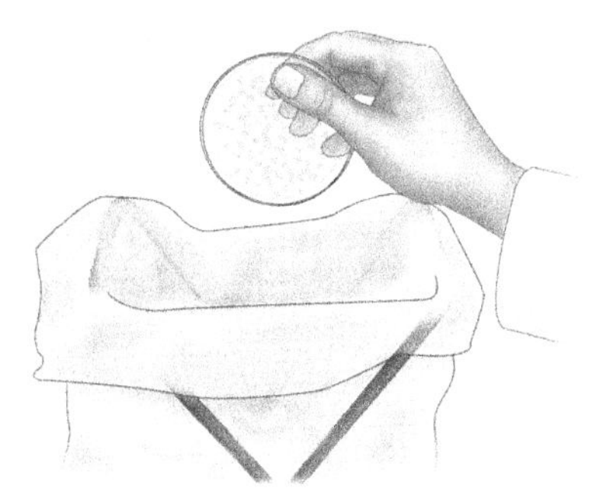

To calculate the concentration of infectious phage in the sample (PFU/mL), use the following equation

$$\frac{\text{PFU}}{\text{mL}} = \frac{\text{number of plaques on plate}}{\text{dilution of sample that was plated}}$$

For example, suppose that 132 plaques were counted on the 10^{-7} plate. When these numbers are used in the equation

$$\frac{\text{PFU}}{\text{mL}} = \frac{132}{10^{-7}} = 132 \times 10^{7}$$

$$= 1.3 \times 10^{9}$$

Because the plaque assay method has at least a 10% margin of error, only two significant numbers appear in the final value.

- *Never use your mouth to draw the bacterial suspension or any dilution into a pipette or tube: use a mechanical pipetting device.*
- *Do not set a contaminated pipette down on the lab bench top. When you have finished using it, place the pipette in the proper place for sterilization.*
- *Place tubes that have been emptied of agar in the designated place for sterilization.*

Troubleshooting

- Use the bottom of the meniscus when you measure the volume of a solution with a pipette.
- Keep the pipette vertical to see the meniscus accurately.
- Know the type of pipette that you are using. Check the tip to distinguish between a serological and a measuring pipette. The volume delivered by a serological pipette is the difference between the location of the meniscus on the volume markings and the very tip of the pipette. A measuring pipette delivers the difference between a zero marking near the tip and the meniscus.
- Be consistent in how you measure volumes for each dilution and how you mix each dilution.

- Keep the base layer agar plates in the incubator until just before you use them. The warm agar will delay the soft agar overlay from gelling long enough for complete spreading.
- Keep the melted soft agar tubes in the hot water bath until you need them.
- Work quickly so that the melted agar in the tube does not solidify before you pour the mixture onto the base layer agar.
- If the soft agar solidifies before it has been completely spread over the base layer agar surface (as evidenced by little piles), discard the plate and repeat for this phage dilution.
- Avoid vigorous swirling of the melted agar to avoid splashing over the edge or onto the lid of the plate.
- Pour the melted agar carefully but quickly into the plate to prevent bubbles.
- Agar will become translucent and lighter in color when it has solidified.
- If the plates are incubated for too long, larger plaques (e.g., phage T1 plaques) formed by some bacteriophage will fuse. Plates can be refrigerated for counting at a later time.
- Check the far edges of the agar for plaques that are easy to miss.
- In a research project, triplicate plates are prepared for each dilution. To calculate PFU/mL, use the average number of colonies for that dilution in the equation.

Testing Antibacterial Medicines: Kirby-Bauer Technique

Purpose:

- To determine the sensitivity of a bacterium to several antibacterial medicines.

The Kirby-Bauer technique is the most common procedure used to determine the effectiveness of antibacterial medicines for killing or inhibiting the growth of an infecting bacteria. Paper disks impregnated with different antibacterial medicines are placed on the surface of a Mueller-Hinton agar plate that has just been heavily inoculated with bacteria. The bacteria will grow into what is called a "lawn." During incubation, as the medicine diffuses out from the disk, the growth or survival of the bacterial cells in the lawn may be affected. A **zone of growth inhibition** may appear around the disk. The diameter of this zone indicates susceptibility or resistance of the tested bacterium to the antibacterial medicine.

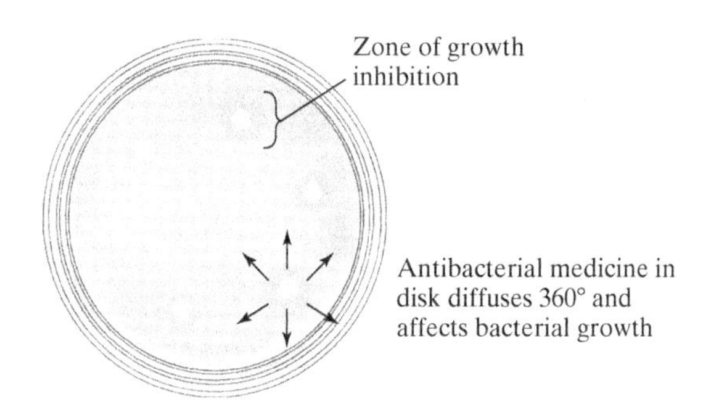

Materials:

- Mueller-Hinton agar plate, 4 mm thick
- antibiotic disk dispenser *or* individual antibiotic disks in cartridges
- forceps
- 95% ethanol in a beaker
- sterile cotton swabs
- ruler, millimeter
- bacterial culture in broth tube, 18-hours old

Procedure:

1. Label the bottom of a Mueller-Hinton agar plate with the name of the bacterium, your name or initials, and the date.

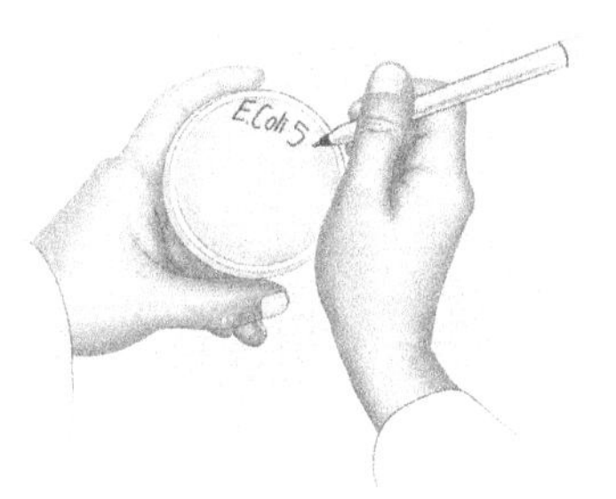

2. Hold the tube of swabs in your non-dominant hand. Curl the little finger of your dominant hand around the cap and remove it. Do not set the cap down.

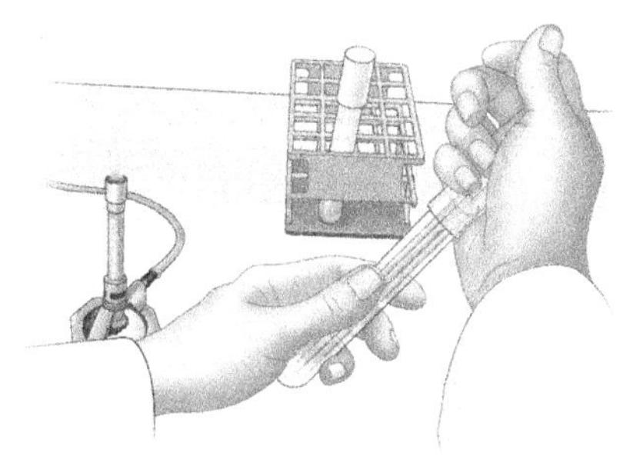

3. Shake the tube so that one of the swabs extends beyond the others. Remove this swab, taking care not to touch the ends of the others. Replace the cap and put the tube down.

4. Pick up the culture tube in your non-dominant hand. Remove the cap by curling your little finger on the hand holding the swab around the cap. Do not set the cap down. Quickly pass the opening of the tube through the flame three times.

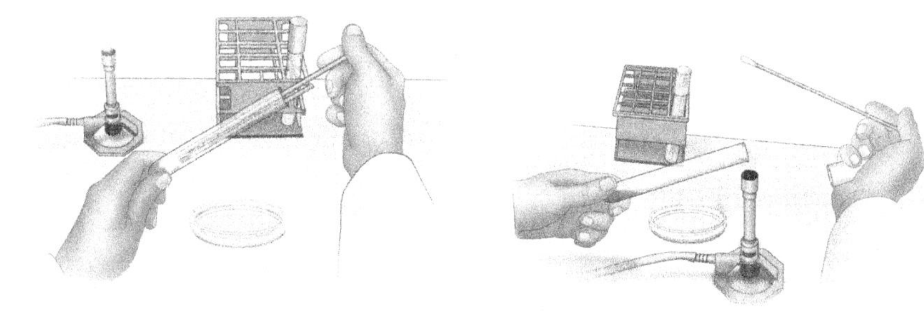

5. Insert the cotton swab into the broth culture. Press the swab against the inside of the tube to remove excess medium.

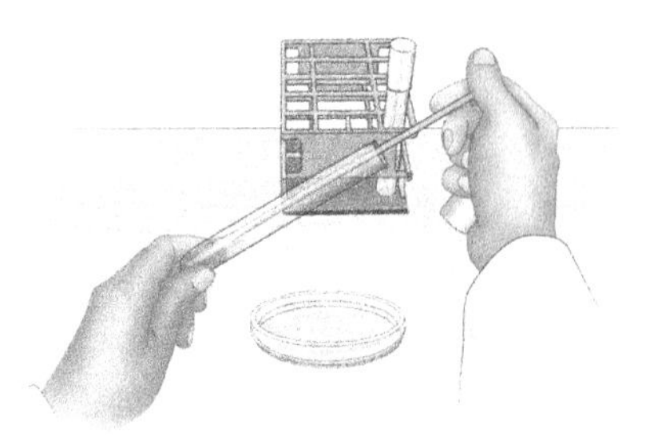

6. Inoculate the entire surface of the agar with a closely spaced back-and-forth motion.

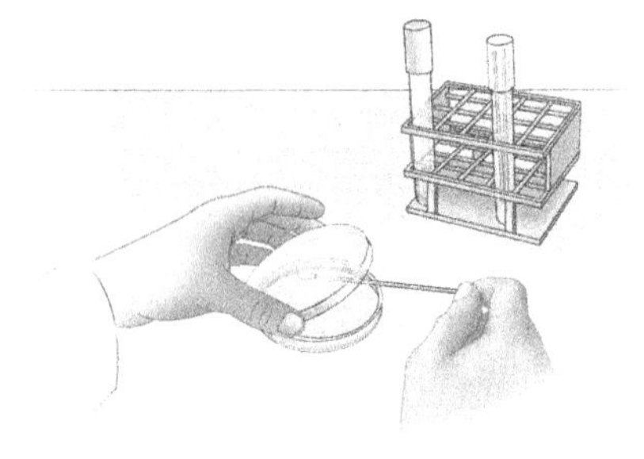

7. Dip the swab into the culture broth again. Inoculate the agar surface in a second direction. Repeat a third time. This technique results in a "lawn" of bacteria on the agar surface.

8. With the plate lid on, allow the inoculated agar surface to dry for 5 minutes.

9. Discard the contaminated swab in the designated place for sterilization.

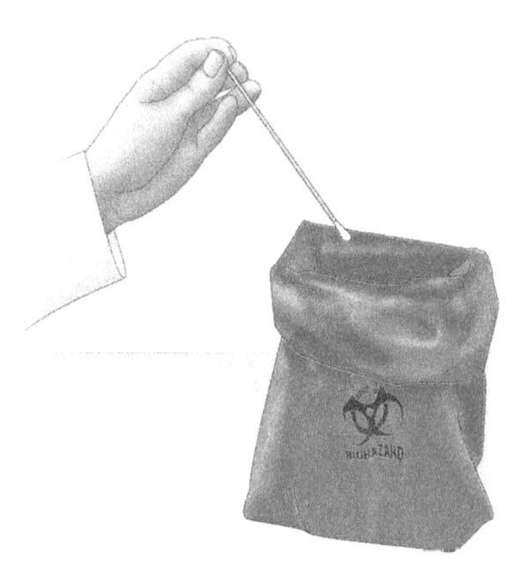

10. If you have a disk dispenser, remove the plate lid and place the dispenser over the open plate. Press the plunger to dispense the disks.

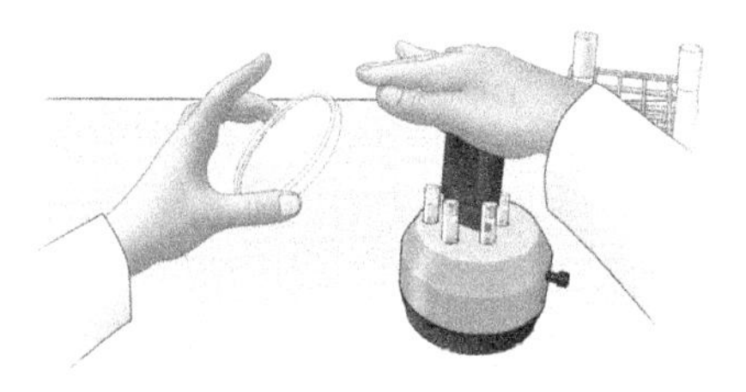

11. If you have disk cartridges, dispense 5 disks evenly spaced and about 20 mm from the plate edge.

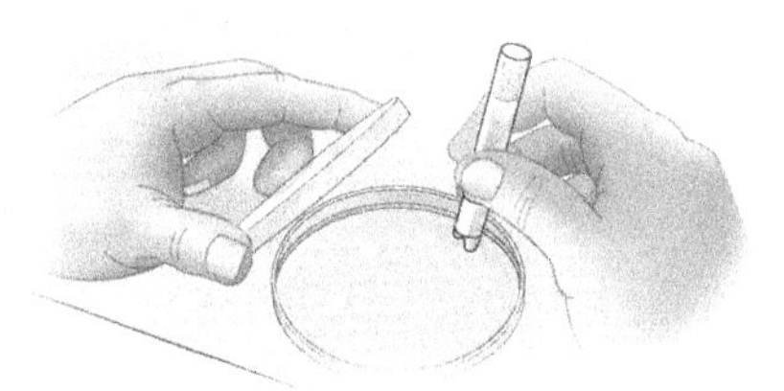

12. Sterilize a forceps by first dipping it in alcohol and then burning off the alcohol.

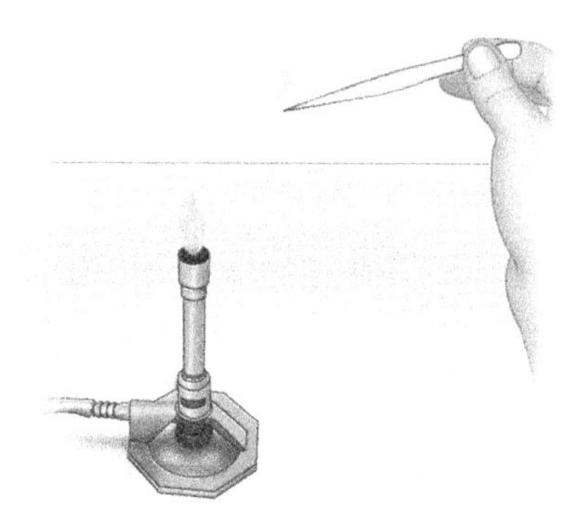

13. Use the sterile forceps to apply light pressure on the disks to prevent them from falling off when the plate is inverted for incubation.

14. Place the plate upside-down in the incubator, and incubate it at 35° C for 16–18 hours.

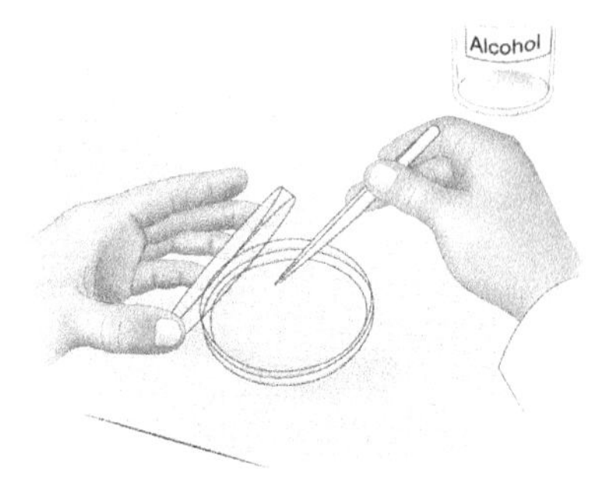

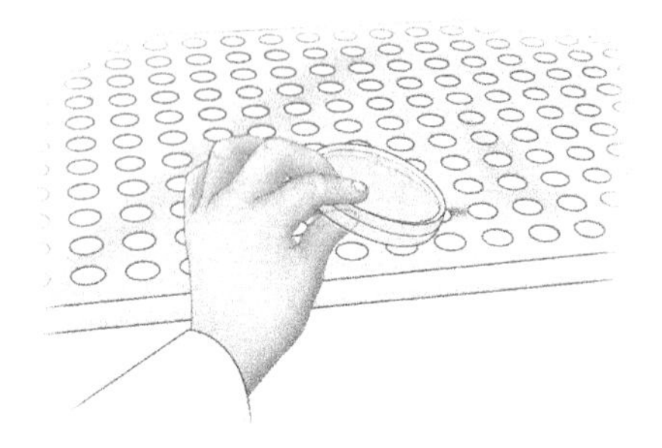

15. After incubation, measure the diameter (in mm) of the zone of growth inhibition for each disk. Place the ruler on underside of the plate. Use the table on the facing page to determine the sensitivity of the bacterium to each antibacterial medicine.

16. Discard the plate in the designated place for sterilization.

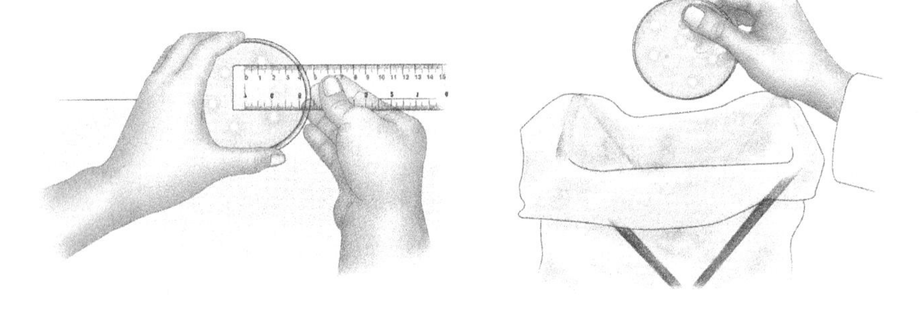

Interpretation of inhibition zones for antibacterial medicines against bacteria

Disk symbol	Antibacterial medicine	Diameters of growth inhibition zones		
		Resistant (≤ mm)	Intermediate (range mm)	Sensitive (≥ mm)
AM 10	Ampicillin 10 μg			
	staphylococci	28	–	29
	enterococci	16	–	17
	Gram-negative enteric bacilli	13	14–16	17
C 30	Chloramphenical 30 μg	12	13–17	18
CC 2	Clindamycin 2 μg	14	15–20	21
CIP 5	Ciprofloxacin 5 μg	15	16–20	21
CR 30	Cephalothin 30 μg	14	15–17	18
DO 30	Doxycycline 30 μg	12	13–15	16
E 15	Erythromycin 15 μg	13	14–22	23
FD 300	Nitrofurantoin 300 μg	14	15–16	17
G300	Sulfisoxazole 300 μg	12	13–16	17
GM 10	Gentamicin 10 μg	12	13–14	15
K 30	Kanamycin 30 μg	13	14–17	18
ME 5	Methicillin 5 μg			
	staphylococci	9	10–13	14
NA 30	Nalidixic acid 30 μg	13	14–18	19
P 10	Penicillin G 10 units			
	staphylococci	28	–	29
	enterococci	14	–	15
	streptococci (not *S. pneumoniae*)	19	20–27	28
RA 5	Rifampin 5 μg	16	17–19	20
S 10	Streptomycin 10 μg	11	12–14	15
SxT 25	Trimethoprim 1.25 μg/ sulfamethoxazole 25.75 μg	10	11–15	16
TE 30	Tetracycline 30 μg	14	15–18	19
TMP 5	Trimethoprim 5 μg	10	11–15	16

Source: Based on data from the Clinical and Laboratory Standards Institute.

- *Position the beaker containing alcohol away from the open flame.*
- *Take care so drops of flaming alcohol do not fall into the beaker or onto flammable materials on the lab bench surface. If the alcohol in the beaker is ignited, cover the beaker immediately to block air and so douse the flames. Know the location of the nearest fire extinguisher.*
- *Do not open the lid to measure zones of growth inhibition; measure on the underside of the plate. This will eliminate possible contamination of the surroundings.*
- *A contaminated cotton swab should not be laid down on the bench top. It must be properly disposed of for sterilization.*

TESTING ANTISEPTICS AND DISINFECTANTS: DISK DIFFUSION TECHNIQUE

Purpose:

- To evaluate the effectiveness of selected antibacterial chemicals in affecting the growth of a particular bacterium.

In this procedure, absorbent paper disks that have been dipped in different chemical solutions are placed on a freshly-prepared "lawn" of bacteria. During incubation, as the chemical diffuses out from the disk, it may affect the growth the bacteria by creating a zone of growth inhibition. Results of this test will indicate only if a particular chemical will adversely affect the growth of a given bacterium.

Materials:

- Mueller-Hinton agar plate, 4 mm thick
- disinfectants and antiseptics in small beakers, identified by letter
- forceps
- 95% ethanol in a 150 mL beaker
- sterile cotton swabs
- ruler, millimeter
- bacterial culture in broth tube, 18-hours old

Procedure:

1. Label the bottom of a Mueller-Hinton agar plate with the name of the bacterium, your name or initials, and the date. Write evenly-spaced capital letters A–D or E about 20 mm from the plate edge.

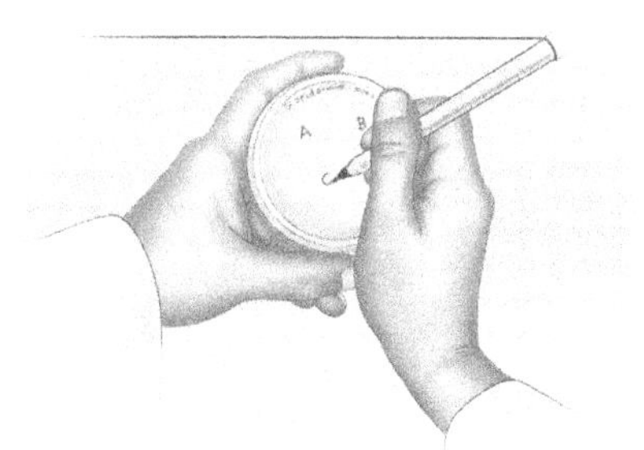

2. Hold the tube of swabs in your non-dominant hand. Curl the little finger of your dominant hand around the cap and remove it. Do not set the cap down.

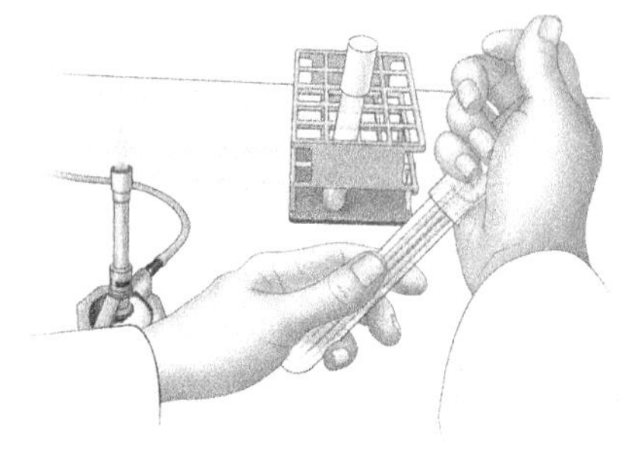

3. Shake the tube so that one of the swabs extends beyond the others. Remove this swab, taking care not to touch the ends of the others. Replace the cap and put the tube down.

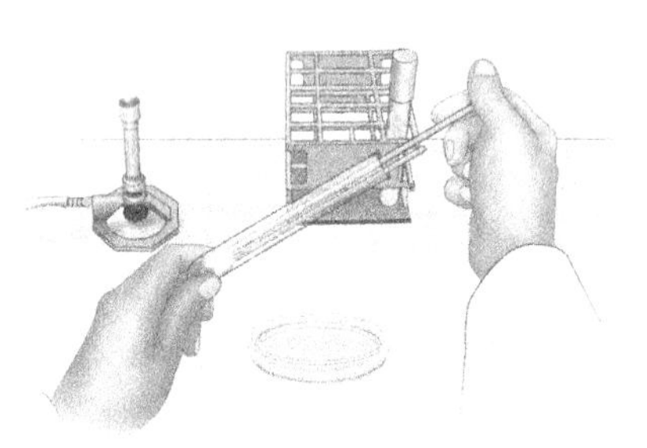

4. Pick up the culture tube in your non-dominant hand. Remove the cap by curling your little finger on the hand holding the swab around the cap. Do not set the cap down. Quickly pass the opening of the tube through the flame three times.

5. Insert the cotton swab into the broth culture. Press the swab against the inside of the tube to remove excess medium.

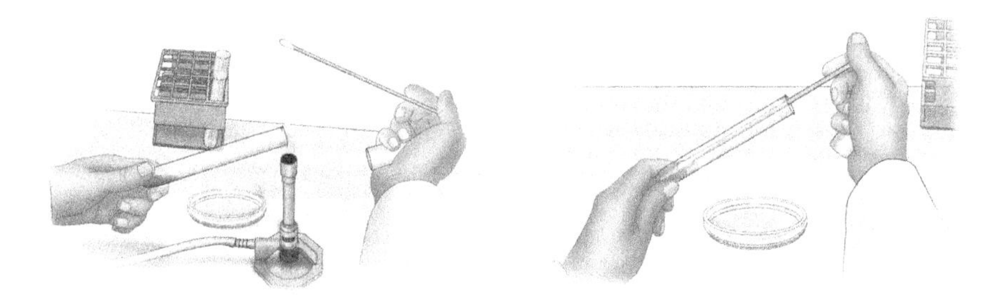

6. Inoculate the entire surface of the agar with a closely spaced back-and-forth motion.

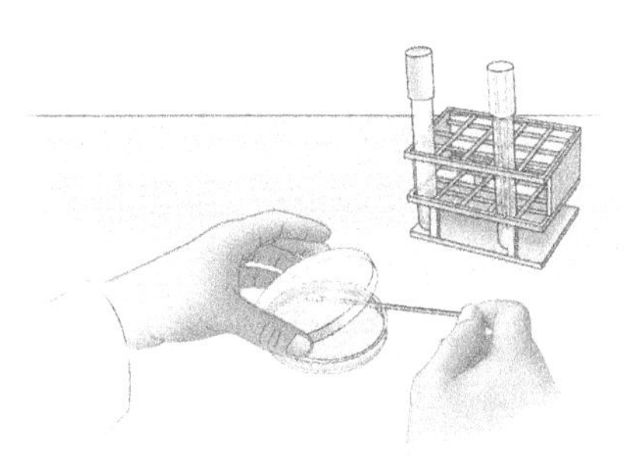

7. Dip the swab into the culture broth again. Inoculate the agar surface in a second direction. Repeat a third time. This technique results in a "lawn" of bacteria on the agar surface.

8. With the plate lid on, allow the inoculated agar surface to dry for 5 minutes.

9. Discard the contaminated swab in the designated place for sterilization.

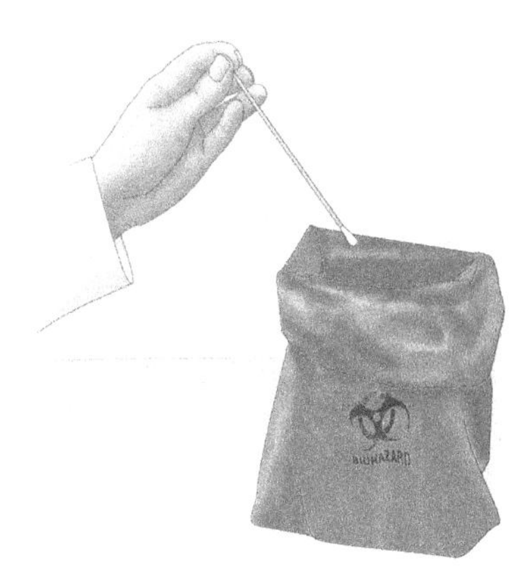

10. Sterilize a forceps by first dipping it in alcohol and then burning off the alcohol.

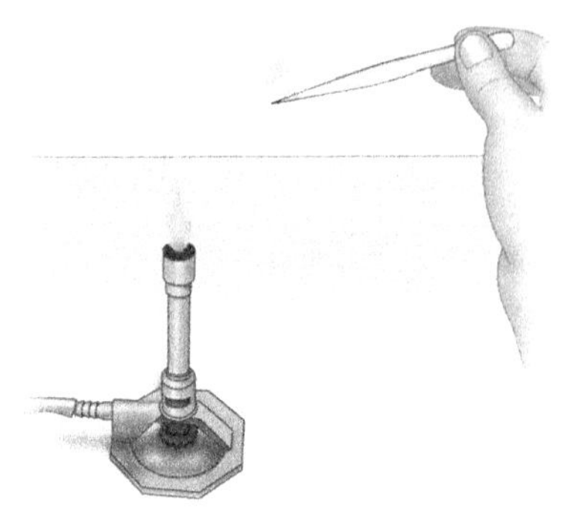

11. Use the sterilized forceps to remove a sterile paper disk from its container.

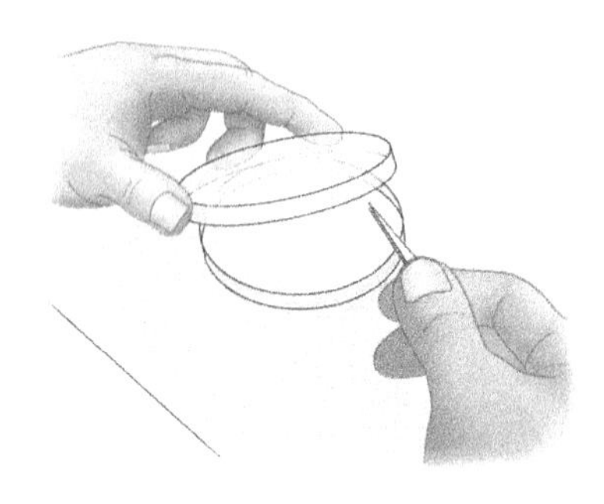

12. Dip the edge of the paper disk into antiseptic/disinfectant A and allow capillary action to saturate the disk.

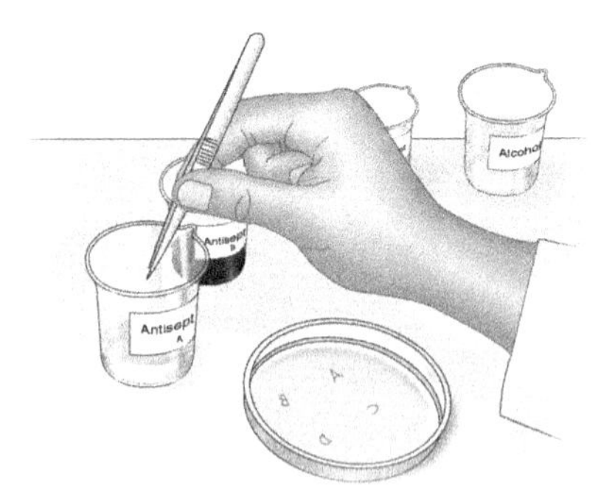

13. Gently place the saturated disk over its letter code. Use the forceps to apply light pressure on the disk to prevent them from falling off when the plate is inverted for incubation.

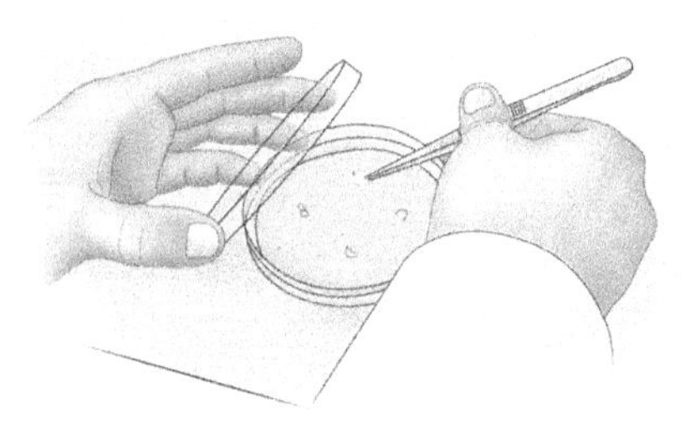

14. Repeat Steps 10–13 for each of the remaining chemical solutions. Flame the forceps for each new solution.

15. Place the plate upside-down in the incubator, and incubate it at 35° C for 18–24 hours.

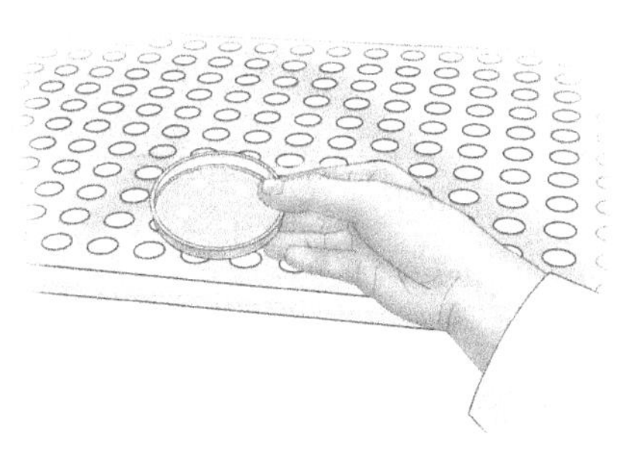

16. After incubation, measure the diameter (in mm) of any zone of growth inhibition around a disk.

17. Discard the plate in the designated place for sterilization.

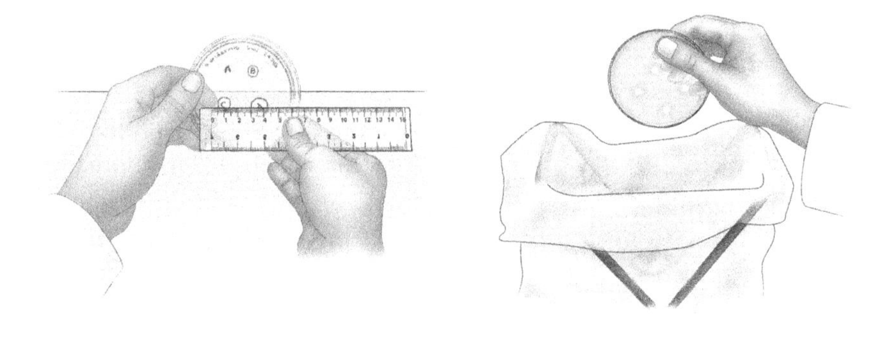

- *Position the beaker containing alcohol away from the open flame.*
- *Take care so drops of flaming alcohol do not fall into the beaker or onto flammable materials on the lab bench surface. If the alcohol in the beaker is ignited, cover the beaker immediately to block air and so douse the flames. Know the location of the nearest fire extinguisher.*
- *Do not open the lid to measure zones of growth inhibition; measure on the underside of the plate. This will eliminate possible contamination of the surroundings.*
- *A contaminated cotton swab should not be laid down on the bench top. It must be properly disposed of for sterilization.*

APPENDIX A

Stains

Acetone-alcohol for Gram stain (alternative decolorizing agent)	acetone	50.0 mL
	ethyl alcohol, 95%	50.0 mL
	Store in tightly sealed bottle.	
Acid-alcohol for acid-fast stain	concentrated HCl	3.0 mL
	ethyl alcohol, 95%	97.0 mL
	• Prepare under a fume hood. • Use a pipettor. Do not mouth pipette. • Add acid to the alcohol slowly.	
Carbolfuchsin for simple stain for acid-fast stain Ziehl-Neelsen (hot method)	Solution A	
	basic fuchsin	0.3 g
	ethyl alcohol	10.0 mL
	Solution B	
	phenol	5.0 g
	distilled water	95.0 mL
	Mix solutions A & B. Let stand in the dark for several days. Filter before use.	
Kinyoun (cold method)	basic fuchsin	4.0 g
	phenol (melted at 60° C)	8.0 mL
	Mix to form a slurry.	
	Then add	
	95% ethyl alcohol	20.0 mL
	Followed by	
	distilled water	100.0 mL
	Mix well. Filter before use. Loses potency after 4 months.	

Crystal violet for simple stain for Gram stain	Solution A crystal violet 95% ethyl alcohol	 2.0 g 20.0 mL
	Solution B ammonium oxalate distilled water	 0.8 g 80.0 mL
	Mix solutions A & B. Let stand for 1 day. Filter before use.	
Gram's iodine for Gram stain	potassium iodide (KI) distilled water	2.0 g 300.0 mL
	Dissolve the KI in the H_2O.	
	iodine (I)	1.0 g
	Grind I with mortar and nestle. Add I to stirring KI solution. Store in a dark bottle.	
Malachite green for endospore stain	malachite green (oxalate salt) distilled water	5.0 g 100.0 mL
Methylene blue for simple stain for acid-fast stain	methylene blue distilled water	1.0 g 100.0 mL
Nigrosin for negative stain	nigrosine distilled water	10.0 g 100.0 mL
	Boil until dissolved.	
	formaldehyde solution	0.5 mL
	Filter.	
Safranin for Gram stain for endospore stain	safranin-O ethyl alcohol, 95%	2.5 g 10.0 mL
	Dissolve in:	
	distilled water	100.0 mL

Reagents

Barritt's reagents for Voges–Proskauer test	Solution A	
	alpha naphthol	6.0 g
	ethyl alcohol, 95%	100.0 mL
	Store in dark bottle at 4° C for no more than one week.	
	 • Wear gloves when you handle this reagent; it is a potential carcinogen.	
	Solution B	
	potassium hydroxide	6.0 g
	distilled water	100.0 mL
Ferric chloride, 10% for phenylalanine deaminase test	ferric chloride	10.0 g
	distilled water	100.0 g
Gram's iodine for detection of starch	See Gram's iodine in *Stains*	
Kovac's reagent for detection of indole	*p*-dimethylamino-benzaldehyde	5.0 g
	amyl (or butyl) alcohol	75.0 mL
	Mix with magnetic stirrer in a fume hood. Gentle heating may be needed. Cool if necessary. Then add:	
	HCl, concentrated	25.0 mL
	Store in a dark bottle at 4° C.	
Methyl red reagent for methyl red test	methyl red	0.1 g
	ethyl alcohol, 95%	300.0 mL
	Add methyl red to ethyl alcohol on magnetic stirrer. Then add distilled water for a final volume of 500 mL.	

Nitrate reduction reagents for detecting nitrite

Solution A

N,N-dimethyl-1-naphthylamine-dihydrochloride	0.6 g
acetic acid, 5N	100.0 mL

(1 part glacial acetic acid: 2.5 parts distilled water)

- Wear gloves when you handle this reagent. It is a potential carcinogen.

Solution B

sulfanilic acid	0.8 g
acetic acid, 5N	100.0 mL

INDEX

A

Acetone-alcohol, preparation of, 218
Acid-alcohol, preparation of, 218
Acid-fast stain
  appearance of bacteria at each step, 104
  Kinyoun (cold) method, 99–103
  Ziehl-Neelsen (hot) method, 99–103
Agar deep culture
  inoculation, 40–41
  preparation of media, 6–7, 10–11
Agar plate
  colony characteristics, 60–62
  inoculation, 42–59
  media preparation, 6–7, 12–16
Agar slant
  growth characteristics, 62
  inoculation, 32–39
  media preparation, 6–7, 10–11
Alpha-hemolytic reaction, 169
Amylase, test for, 121–122
Amino acid decarboxylase test, 148–149
Anaerobic culture techniques
  candle jar, 73
  GasPak jar, 71–72
  thioglycollate broth, 70
Antibiotics, sensitivity testing, 206–212
Antiseptics, testing effectiveness of, 213–217
Aseptic procedures
  preparation of work area, 28
  transferring cultures, 29–41
Autoclave, use of, 17–18

B

Bacterial colonies
  characteristic features of, 60–61
Bacterial cultures. *See* Cultures, bacterial
Bacti-Cinerator, use of, 27
Bacteriophage, determining titer, 198–205
Barritt's reagents, preparation of, 220
Beta-hemolytic reaction, 169
Blood agar plate, 168–169
Blood and body fluid precautions, 5
Broth culture
  growth characteristics, 62
  inoculation, 29–31
  preparation of media, 6–9
Bunsen burner, adjusting, 23–26
Butanediol fermentation
  test for, 135–136

C

Candle jar, use of, 73
Capsule stain, 109–111
Carbohydrates, fermentation of
  butanediol pathway, 135–136
  Durham tube, 131–132
  mixed, 133–134
Carbolfuchsin, preparation of, 218
Casein hydrolysis test, 123–124
Catalase test, 150–151
Citrate utilization test, 137–138
Coagulase (tube) test, 166–167
Cold method (Kinyoun) acid-fast staining, 99–103
Colony-forming units (CFUs), 183, 189
Counterstain
  acid-fast stain, 104
  endospore stain, 116
  Gram stain, 98
Counting bacteria
  direct (Petroff-Hausser chamber), 178–182
  indirect (turbidimetry), 191–197
Cultures, bacterial
  reserve, 63
  storing, 63–65
  transferring, 29–39
  working, 63
Crystal violet
  capsule stain, 109
  Gram stain, 94, 98
  preparation of, 219
  simple stain, 92

D

Decolorizing agent
 acid-fast stain, 101, 104
 Gram stain, 95, 98
Differential media
 blood agar, 168
 eosin methylene blue (EMB) agar, 170
 mannitol salt agar, 172
 MacConkey agar, 174
 Simmons citrate agar, 137-138
Dilution procedures, 181, 184–187
Disk diffusion method
 for testing antiseptics and disinfectants, 213–217
Disinfectants, testing effectiveness of, 213–217
DNA hydrolysis test, 129–130
Dry heat, sterilization by, 18
Durham tube test, 131

E

Endospore stain
 appearance of bacteria at each step, 116
 procedure, 112–115
Eosin methylene blue (EMB) agar, 170–171

F

Fat hydrolysis test, 127–128
Fermentation, 131
Filtration, sterilization by, 19–22
Freezing, preservation of cultures, 63–64

G

Gelatin hydrolysis test, 125–126
Gram stain
 appearance of bacteria at each step, 98
 procedure, 94–97
Gram's iodine, preparation of, 219
Growth medium, preparation
 agar deep tube, 6–7, 10–11
 agar plate, 6–7, 12–16
 agar slant tube, 6–7, 10–11
 broth, 6–7, 8–9

H

Hanging-drop slide, 84–86
Hemolytic reactions, 168–169
Hot air oven, sterilization by, 18
Hot method (Ziehl-Neelsen) acid-fast staining, 99–103
Hydrogen sulfide production test, 144–145

I

Identification of bacteria
 biochemical tests, 121–177
 staining procedures
  acid-fast stain, 99–104
  endospore stain, 112–116
  Gram stain, 94–98
Indole production (tryptophan degradation) test, 142–143
Inoculation of culture media
 agar deep tube, 40–41
 agar plate, 42–59
 agar slant tube, 32–39
 broth, 29–31
Isolation of bacteria
 pour plate, 48–55
 spread plate, 56–59
 streak plate, 42–47

K

Kinyoun (cold method) acid-fast staining, 99–103
Kirby-Bauer method of assessing antibiotic effectiveness, 206–212
Kovac's reagent, preparation of, 220

L

Lab safety, 1–4
"Lawn" of bacteria, preparation of, 205, 207–208, 214–215
Light microscope
 parts, 74
 steps for use, 75–83
Litmus milk reactions, 161–163

M

MacConkey agar, 174–175
Malachite green, preparation of, 219
Mannitol salt agar, 172–173
Measurement, bacterial cell size, 178–182
Media. *See* Growth medium, preparation

Methylene blue
acid-fast stain, 99, 101, 104
preparation of, 219
simple stain, 92
Methyl red
reagent, preparation of, 220
test, 133–134
Microscope. *See* Light microscope
Mixed acid fermentation, test for, 133–134
Motility testing, 164–165
MR-VR testing, 133–136
Mueller-Hinton agar, 206, 213

N

Negative stain, 105–108
Nigrosin
capsule stain, 109
negative stain, 105
preparation of, 219
Nitrate reduction
reagent, preparation of, 221
test, 154–156
Non–acid-fast bacteria, appearance of, 104

O

Ocular micrometer, calibration of, 117–119
Oxidase test, 152–153
Oxidation–fermentation (OF) test, 154–156

P

Petri plate
preparation of, 12–16
streaking, 42–47
Petroff-Hausser chamber for counting cells, 178–182
pH, adjusting, 7
Phenylalanine deamination test, 146–147
ferric chloride, preparation of, 220
Phenylethyl alcohol (PEA) agar, 176–177
Pipette, use of, 2, 184–187
Pipettor
an inexpensive, 183
use of, 8
Plaques, bacteriophage, 198
Plaque-forming unit (PFU), 198
Pour plate method, 48–57
Preservation of cultures, 63–64, 67–69
Pure culture techniques, 42–59

R

Reagents, preparation of, 220–221

S

Safety
blood and body fluid precautions, 5
ethyl alcohol and gas burner, 56
laboratory, 1–4
Safranin
Gram stain, 94
preparation of, 219
Selective media
eosin methylene blue (EMB) agar, 170
mannitol salt agar, 172
MacConkey agar, 174
phenylethyl alcohol (PEA) agar, 176
Serial dilution, preparation, 184–187
SIM (sulfide-indole-motility) agar, 142
Simmons citrate agar, 137–138
Simple stain, 87–91
Slants, preparation of, 6–7, 10–11
Smear, preparation of, 87–91
Spectrophotometer, use of, 191, 193–196
Spore stain, 112–116
Spread plate method, 56–59
Stab technique
for agar deep cultures, 40–41
Stage micrometer, 117
Stains, preparation of, 218–219
Staining techniques
acid-fast stain, 99–103
capsule stain, 109–111
endospore stain, 112–115
Gram stain, 94–97
negative stain, 105–108
simple stain, 92–93
Standard plate count, 183–190
Staphylococci
coagulase, 166–167
mannitol salt agar, 172–173
Starch hydrolysis test, 121–122

Sterilization
autoclave, 17–18
Bacti-Cinerator, 27
dry air, 18
filtration, 19–22
Stock culture, 63
Storage of bacteria, 63–65
Streak plate method, 42–47
Syringe membrane filter sterilization, 19–20

T

Thioglycollate broth, for anaerobes, 70
Titer of bacteriophage, 198–205
Triglyceride hydrolysis test, 127–128
Triple sugar iron (TSI) agar test, 157–158
Tryptophan degradation (indole production) test, 142–143
Turbidimetry (indirect counting of bacteria), 191–197

U

Urea hydrolysis test, 159–160

V

Voges-Proskauer test, 135–136

Z

Ziehl-Neelsen (hot method) acid-fast staining, 99–103
Zone of inhibition
antibiotics, evaluation of, 206–212
antiseptics and disinfectants, evaluation of, 213–217